Jonas Cicenas

"ErbB2 signaling in breast cancer"

Jonas Cicenas

"ErbB2 signaling in breast cancer"
"the role of ErbB, Akt and ShcA phosphorylation"

Südwestdeutscher Verlag für Hochschulschriften

Impressum/Imprint (nur für Deutschland/only for Germany)
Bibliografische Information der Deutschen Nationalbibliothek: Die Deutsche Nationalbibliothek verzeichnet diese Publikation in der Deutschen Nationalbibliografie; detaillierte bibliografische Daten sind im Internet über http://dnb.d-nb.de abrufbar.
Alle in diesem Buch genannten Marken und Produktnamen unterliegen warenzeichen-, marken- oder patentrechtlichem Schutz bzw. sind Warenzeichen oder eingetragene Warenzeichen der jeweiligen Inhaber. Die Wiedergabe von Marken, Produktnamen, Gebrauchsnamen, Handelsnamen, Warenbezeichnungen u.s.w. in diesem Werk berechtigt auch ohne besondere Kennzeichnung nicht zu der Annahme, dass solche Namen im Sinne der Warenzeichen- und Markenschutzgesetzgebung als frei zu betrachten wären und daher von jedermann benutzt werden dürften.

Verlag: Südwestdeutscher Verlag für Hochschulschriften GmbH & Co. KG
Dudweiler Landstr. 99, 66123 Saarbrücken, Deutschland
Telefon +49 681 37 20 271-1, Telefax +49 681 37 20 271-0
Email: info@svh-verlag.de

Approved by: Basel, U, Diss., 2004

Herstellung in Deutschland:
Schaltungsdienst Lange o.H.G., Berlin
Books on Demand GmbH, Norderstedt
Reha GmbH, Saarbrücken
Amazon Distribution GmbH, Leipzig
ISBN: 978-3-8381-2971-6

Imprint (only for USA, GB)
Bibliographic information published by the Deutsche Nationalbibliothek: The Deutsche Nationalbibliothek lists this publication in the Deutsche Nationalbibliografie; detailed bibliographic data are available in the Internet at http://dnb.d-nb.de.
Any brand names and product names mentioned in this book are subject to trademark, brand or patent protection and are trademarks or registered trademarks of their respective holders. The use of brand names, product names, common names, trade names, product descriptions etc. even without a particular marking in this works is in no way to be construed to mean that such names may be regarded as unrestricted in respect of trademark and brand protection legislation and could thus be used by anyone.

Publisher: Südwestdeutscher Verlag für Hochschulschriften GmbH & Co. KG
Dudweiler Landstr. 99, 66123 Saarbrücken, Germany
Phone +49 681 37 20 271-1, Fax +49 681 37 20 271-0
Email: info@svh-verlag.de

Printed in the U.S.A.
Printed in the U.K. by (see last page)
ISBN: 978-3-8381-2971-6

Copyright © 2011 by the author and Südwestdeutscher Verlag für Hochschulschriften GmbH & Co. KG and licensors
All rights reserved. Saarbrücken 2011

ErbB2 Signaling in Breast Cancer: the role of ErbB, Akt and ShcA phosphorylation

Inauguraldissertation

zur

Erlangung der Würde eines Doktors der Philosophie

vorgelegt der

Philosophisch-Naturwissenschaftlichen Fakultät

Der Universität Basel

von

Jonas Cicenas

aus Klaipeda (Litauen)

Basel, 2004

Genehmigt von der Philosophisch-Naturwissenschatftlichen Fakultät
auf Antrag der Proffessoren
Nancy Hynes, Urs Eppenberger, Gerhard Christofori und Dr. Willy Kueng

Basel, den 16.11.2004

Dekan, Prof. Dr. Hans-Jakob Wirz

Acknowledgements

I am most grateful to the members of my thesis committee who made this work possible, in particular Professor Urs Eppenberger for giving me the opportunity to be a member of his research group and to Dr. Willy Kueng for his help with and suggestions. I would also like to thank Prof. Nancy Hynes and Prof. Gerhard Christofori for being my committee members.

Special thanks to Dr. Serenella Eppenberger-Castori for her help with statistical analysis and support, which I received while working at Stiftung Tumorbank Basel.

I would also like to thank the members of "Molecular Tumor Biology" group: Heidi Bodmer and Francoise David for the nice working atmosphere and technical help. Many thanks to Dr. Patrick Urban for the help with statistical analysis and for the DNA microarray data. Lots of thanks to members of OncoScore AG: Dr. Vincent Vuaroqueaux and Dr. Martin Labuhn for the support, suggestions and discussions as well as the Q-RT-PCR data and Sabine Ehret for the technical support.

I am also grateful to the Prof. David M. Terrian for guiding me through my first PhD project "PKCε in Prostate Cancer" and Dr. Daqing Wu for his input in the project.

Thanks to Dr. Mindaugas Valius for getting me interested in Signal Transduction.

At last, but not least, I would like to thank my family, particularly my wife Ernesta, for her patience and encouragement.

Table of contents

Summary..5

1. Introduction

1.1. Breast cancer..8
1.2. Signaling by Receptor Tyrosine Kinases and EGFR family......18
1.3. Shc adaptor proteins..28
1.4. Akt Signaling..31

2. Experimental Procedures

2.1. Materials 37
2.2. Cell culture..39
2.3. Tumor extract preparation for immunoassays.....................39
2.4. Measurement of ER and PgR levels...................................40
2.5. Immunoassay of total ErbB2 receptor levels.......................40
2.6. Immunoassay of P-Y1248 ErbB2 level 40
2.7. Immunoassays of pan-Y, S and T phosphorylation of ErbB2.....41
2.8. Immunoassay of P-Akt level...42
2.9. Immunoassay of pan-Y and S phosphorylation of ShcA..........43
2.10. Isolation of mRNA..44
2.11. cDNA synthesis..44
2.12. Primer design..44
2.13. Quantitative real-time RT-PCR......................................45
2.14. EGF binding assay for ERGFR quantification....................45
2.15. Sample preparation for SDS-PAGE..................................46
2.16. SDS-PAGE...46
2.17. Western-Blotting..46
2.18. ECL detection..46
2.19. Computing...46
2.20. Statistics..46

3. Immunoassay development
3.1. Introduction..48
3.2. Results..49

4. The role of Y1248 ErbB2 phosphorylation in primary breast cancer
4.1. Introduction..55
4.2. Results..56
4.3. Discussion...64

5. The role of Akt phosphorylation in primary breast cancer
5.1. Introduction..66
5.2. Results..67
5.3. Discussion...75

6. Analysis of ErB2, ShcA and Akt phosphorylation in primary breast cancer
6.1. Introduction..78
6.2. Results..79
6.3. Discussion...88

7. Discussion and perspectives......................90
List of abbreviations............................93
References..96
Appendix
a.1. Assay schemes..125
a.2. PKCepsilon in prostate cancer...129

Summary

Breast cancer is the most common malignancy in women and is estimated to account for more than 200,000 new cancer cases in the United States in the year 2002. It now represents the second leading cause of death (40,000) from cancer in women. Although the number of new breast cancer cases has been increasing, the death rate has been steadily decreasing. This trend may be due to earlier diagnosis, and/or increased survival resulting from the use of adjuvant therapy.

Clinical outcome is affected by prognostic predictive factors. Prognostic factors are associated with either the metastatic or the growth potential of the primary tumor, while predictive factors are associated with the relative sensitivity and/or resistance to specific therapies. Routinely available prognostic indicators include tumor size, type, and grade, axillary lymph node status, estrogen and progesterone receptor status. Estrogen receptor status and progesterone receptor status also serve as predictive factors for expected response to hormone therapy. Many other molecular markers are being investigated for their clinical usefulness. One of the major molecular prognostic and predictive markers in breast cancer is the amplification status of the proto-oncogene ErbB2 (HER-2/ *neu*).

The ErbB2 proto-oncogene is a component of a four-member family of closely related growth factor receptors that includes the epidermal growth factor receptor (ErbB1/HER1), ErbB3 (HER3), and ErbB4 (HER4). The human gene is located on chromosome 17q21 and encodes a 185-kDa protein with tyrosine kinase activity that is also known by the designation p185. Structurally, the protein has extracellular, transmembrane, and a cytoplasmic domain, the latter of which contains the tyrosine kinase domain and shares significant homology, although is distinct, from EGFR. Under normal circumstances, low levels of ErbB2 expression are detectable immunohistochemically in a variety of fetal and adult epithelial cells throughout the gastrointestinal, respiratory, and genitourinary tracts. Amplification of the ErbB2 proto-oncogene or overexpression of the p185 protein, which generally correlate with each other, has been identified in 10% to 34% of breast cancers as well as in gastrointestinal, pulmonary, and genitourinary tumors. The mechanism by which overexpressed ErbB2 leads to a neoplastic phenotype occurs by activation of several different signaling pathways that lead to gene activation, ultimately resulting in cell proliferation. Although the mechanism of activation of ErbB2 has not been completely elucidated, it is thought to involve the formation of heterodimers with other members of the epidermal growth factor family of receptors or spontaneous homodimerization.

This study was designed to compare the prognostic value of phosphorylated ErbB2 in well characterized primary breast cancer samples. Seventy primary breast cancers with a median of 45 months of follow-up were analyzed for quantitative levels of phosphorylated ErbB2 using new sensitive chemiluminescence-linked immunoassay (CLISA). Phosphorylated ErbB2 data were compared with clinical, histological and outcome variables as well as quantitative mRNA and protein expression levels of ErbB family members. ErbB2 – overexpressing tumors contained significantly more phosphorylated ErbB2, however PY1248 could be detected in some of low ErbB2 expression tumors. ErbB2 phosphorylation was correlated with disease free and overall survival and reduced estrogen receptor and progesterone receptor contents. Comparison of ErbB family expression on mRNA level with ErbB2 phosphorylation revealed significant correlation with ErbB2 and EGFR but inverse correlation with ErbB3 and ErbB4. Similar correlations were found also with respect to protein expression levels of these factors.

We have also investigated total (pan) tyrosine, serine and threonine phosphorylated ErbB2 in 153 breast cancer samples by two-site CLISA assays. Serine and threonine phosphorylated ErbB2 could be detected only in low ErbB2 – expressing tumors, no serine and threonine phosphorylation was detectable in ErbB2 overexpressing tumors. As in case of PY1248, ErbB2 – overexpressing tumors contained significantly more tyrosine phosphorylated ErbB2, but ErbB2 tyrosine phosphorylation was detectable in some of low ErbB2 expression tumors as well. Due to the fact that tumors we selected for this study were mostly aggressive tumors, it was impossible to analyze the prognostic value of ErbB2 phosphorylations.

Akt1, Akt2 and Akt3 kinases are involved in the signal transduction pathway downstream of receptor tyrosine kinases via phosphoinosytol-3-kinase, influencing cell growth, proliferation and survival. Akt2 overexpression and amplification have been described in breast, ovarian and pancreatic cancers. In this study we measured the quantitative expression levels of total phosphorylated (P-S473) Akt (Akt1/2/3) by means of a two-site CLISA on cytosol extracts obtained from 156 primary breast cancer tissue samples. We aimed to clarify the prognostic significance of activated Akt in primary breast cancer in association with other tumor biomarkers. Akt phosphorylation was not associated with the nodal status and the ErbB2 expression. Only very high expression levels of P-Akt correlated with poor prognosis. More importantly, the prognostic value of P-Akt expression increased in ErbB2 overexpressing subset of patients. In addition, P-Akt was found to be associated with mRNA expression levels of several proliferation markers, such as thymidylate synthase, thymidine kinase 1, survivin, topoisomerase II alpha and transcription factor E2F, measured by quantitative real-time PCR (Q-RT-PCR).

Shc adapter/docking proteins are an important component of receptor tyrosine kinase signaling pathways because they are involved in transducing the activation signals from receptor or cytoplasmic tyrosine kinases to downstream signaling cascades. At least three genes, *shcA*, *shcB*, and *shcC*, are known to encode Shc proteins. ShcA has been found to be phosphorylated rapidly and efficiently by all tyrosine kinases tested to date. These phosphorylation sites have been mapped to Y339, Y240, and Y317. In addition to tyrosine phosphorylation, ShcA can also be phosphorylated at serine/threonine residues.

We have investigated pan- tyrosine, serine and threonine phosphorylated ShcA in 153 breast cancer samples by two-site CLISA assays. P-ShcA was found to be weekly associated with PT ErbB2 levels and weekly inversely correlated with P-Akt levels. A very good correlation was found between PS ShcA and PY SchA.

Since it was the same collective of tumors, as the one used for ErbB2 pan-S, T and Y phosphorylation assessment, it was also impossible to analyze the prognostic value of phospho-SchA.

1. Introduction

1.1. Breast cancer.

1.1.1. Epidemiology. Breast cancer is the most predominant tumor among women in Western countries. The incidence rate is 70-100 cases per 100,000 women a year and the mortality rate 20-30 deaths per 100,000 women a year in Western countries and up to 5 fold less in Eastern Asia and Africa (1, 2). There were more than 200,000 estimated new cases in the United States alone, in 2002 and 40,000 estimated breast cancer deaths (3). Recognized risk factors for breast cancer are age, increased hormone exposure and genetic predisposition. Brest cancer is age-dependent, e.g. incidence in North America and Europe is about 2.5% by the age of 55, 5% by the age of 65 and 7.5% by the age of 75(2). Increased hormone exposure, such as early menarche, late menopause, oral contraceptics, hormonal therapy together with alcohol consumption and obesity is associated with increased risk. Breast feeding, early first pregnancy and physical exercise on the other hand are associated with a reduced risk (1, 2). The majority of breast cancers arise sporadically. However, family history is responsible for about 2-5% of breast cancers. Genes involved in hereditary forms of breast cancer include BRCA1, BRCA2, P53, STK11/LKB1, PTEN and ATM (1, 2).

1.1.2. Biology of breast cancer. The mammary gland is a highly differentiated organ that is responsible for providing nutrition to the progeny. Mammary development starts during embryogenesis; in humans, males and females have a similar rudimentary mammary gland at birth. Later mammary development is initiated with the beginning of female puberty and is dependent on the high levels of estrogen produced by the ovary, as well as levels of progesterone. After puberty, the mammary gland undergoes cycles of growth and involution, regulated by the menstrual cycle, cycles of pregnancy and lactation. Histologically, the mammary gland consists of a rudimentary branching duct system lying in a fat pad (Fig. 1). Post-pubertal development results in cyclical increases in ductal branching, resulting in a ductal tree that fills the fat pad. During pregnancy, further branching and end-bud development lead to an appearance that is like bunches of "grapes". After weaning, mammary-gland regression to a pre-pregnancy like state is manifested by apoptotic processes. The ductal structure consists of a continuous layer of epithelial cells responsible for milk synthesis and release into the lumen. A second layer of myoepithelial cells contacts the basement membrane. The two cell layers, together with fibroblasts surrounding, form the basis of the ducts (4, 5). Epithelial cells are sites of estradiol action in the breast, according to immunohistochemical analysis demonstrating that the epithelial cells estrogen receptor (ER) (6). According to the literature the luminal epithelial cells that are responsible for most breast tumors. The pathway to breast cancer development is not clear. There is some evidence, although inconclusive, to indicate that it might begin with hyperproliferation of the epithelial cells, progressing through a preneoplastic phase called ductal carcinoma *in situ* (DCIS), which is bounded by the basement membrane, to invasive breast cancer, in which the basement membrane has been breached. About 15–25% of epithelial cells are ER-positive in the normal resting breast, although the number of ER-positive cells changes throughout the menstrual cycle. Interestingly, estrogen stimulated proliferation occurs in ER-negative cells that surround the ER-positive luminal epithelial cells (7, 8). This has led to the suggestion that ER-positive epithelial cells promote proliferation of surrounding ER-negative cells, probably through secretion of paracrine factors. By contrast, proliferation of ER positive epithelial cells in breast tumors is estrogen regulated in the majority of cases. The sequence of events that enable ER-positive cells to be transformed from nondividing cells to a state in which their proliferation is estradiol dependent is at present unclear.

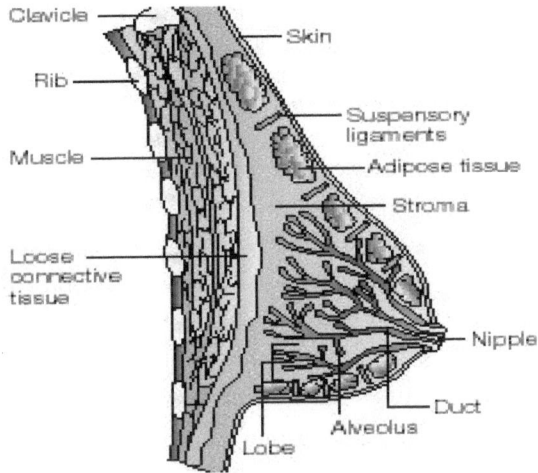

Figure 1. Anatomy of the human mammary gland. Each mammary gland contains 15–20 lobes, each lobe containing branched ducts that drain into the nipple (Nat Rev Cancer. 2002 Feb;2(2):101-12).

There is considerable evidence that links increased breast cancer risk with prolonged exposure to estrogens. This includes the increased risk associated with early menarche, late first full-term pregnancy and late menopause. Oral contraceptives and estrogen-replacement therapy have also been implicated in breast cancer risk. In addition, dietary and environmental agents that can act as estrogens have been linked to breast cancer risk, although their true involvement in breast cancer remains to be established (9, 10). It is possible that the breast-cancer-promoting effects of estrogens arise simply from their proliferative effects on the breast. It is definitely clear that in large amount of cases estradiol promotes breast cancer progression by stimulating malignant cell proliferation. This is further confirmed by the apparent correlation between ER positivity in breast tumors and their response to endocrine therapy. Moreover, ER expression in benign breast epithelium is higher in patients with breast cancer than it is in women who do not have breast cancer (11), demonstrating that the ER is involved in early events in breast cancer. This might reflect deregulated control of ER expression in preneoplastic breast cancer cells, although the primary mechanisms are not defined. Nevertheless, even though only 15–25% of normal breast epithelial cells are ER-positive, two-thirds of breast cancers are ER-positive and approx. 50% of these respond to endocrine therapy (12).

1.1.3. Diagnosis and treatment. The mammography, ultrasound, fine needle aspirations, needle (core) biopsies, blood (serum) tests and excision biopsies may be used in the diagnosis of cancer of the breast (13-15). In addition to detecting breast cancer, or confirming the initial diagnosis, these tests are also useful in showing the extent (stage and grade) of the cancer. The stage of a cancer is a term used to describe its size and whether it has spread beyond its original site (16-17). Knowing the extent of the cancer helps to decide on the most appropriate treatment modalities. A commonly used staging system is described below:

Ductal carcinoma in situ (DCIS): DCIS is when the breast cancer cells are completely contained within the breast ducts, and have not spread into the surrounding breast tissue. This may also be referred to as non-invasive or intraductal cancer, as the cancer cells have not yet spread into the surrounding breast tissue and so usually have not spread into any other part of the body. DCIS is almost always completely curable with treatment.

Lobular carcinoma in situ (LCIS) means that cell changes are found in the lining of the lobules of the breast. It can be present in both breasts. It is also referred to as non-invasive cancer as it has not spread into the surrounding breast tissue.

Invasive breast cancer is divided into 4 stages, from small and localized (stage 1) to metastatic cancer (stage 4). Stage 1 tumors: these measure less than two centimeters. The lymph glands in the armpit are not affected and there are no signs that the cancer has spread elsewhere in the body. Stage 2 tumors: they measure between two and five centimeters, or the lymph glands in the armpit are affected, or both. However, there are no signs that the cancer has spread further. Stage 3 tumors: these are larger than five centimeters and may be attached to surrounding structures such as the muscle or skin. The lymph glands are usually affected, but there are no signs that the cancer has spread beyond the breast or the lymph glands in the armpit. Stage 4 tumors: these are of any size, but the lymph glands are usually affected and the cancer has spread to other parts of the body.

Grading refers to the appearance of the malignant cells, the differentiation status (18). The grade gives an idea of how rapidly the cancer may develop. There are three grades:
1. grade 1 (low grade),
2. grade 2 (moderate grade) and
3. grade 3 (high grade).

Low grade means that the cancer cells look differentiated, like the normal cells of the breast. They are usually slowly proliferating and are less likely to spread. In high grade tumors the cells look undifferentiated. They are likely to proliferate faster and are more likely to spread.

The treatment of breast cancer depends on several factors, such as: the stage of the disease, age of the patient, menopausal status, the size of the tumor, the grade, whether they express receptors for ER and/or PgR, or other proteins such as ErbB2.

In the earliest stages surgery may be all that is needed, but surgery is often followed by radiotherapy to the remaining breast tissue to make sure that any remaining tumor cells are destroyed, particularly if only part of the breast has been removed.

Because there is a risk of cancer cells having escaped and being present elsewhere in the body, additional drug treatment (*adjuvant therapy*) might be necessary. Adjuvant therapy may consist of hormonal therapy or chemotherapy or both.

If the cancer has spread, it is usually treated with drugs (hormonal therapy, chemotherapy or monoclonal antibody therapy). Whether hormonal therapies or chemotherapy are used will depend on the organ that the cancer has metastasized to, how much time has passed since the original surgery and whether or not the cancer cells have receptors for particular hormones or proteins on their surface. Several different hormonal therapies and many different types of chemotherapies can be used.

Chemotherapies or hormonal therapies are sometimes used to shrink a large breast cancer before surgery. When the treatments are given before surgery it is known as *neo-adjuvant therapy*.

Surgery. Several types of surgery are performed in order to remove breast tumors, such as: lumpectomy (wide local excision), quadrantectomy (segmental excision) and mastectomy (19).

For many patients a mastectomy may not be necessary. It is now often possible to just remove the area of cancer and some of the healthy surrounding tissue, and then give radiotherapy to the remaining breast tissue. This is known as breast conserving therapy. Research has shown that in early breast cancer, lumpectomy followed by radiotherapy is as effective at curing the cancer as mastectomy (19).

As part of any surgery for breast cancer the surgeon will usually remove lymph glands from under an arm on the same side of the body. The lymph glands are examined to check whether any cancer cells have spread into them from the breast.

Radiotherapy is most often used after surgery for breast cancer, but is sometimes used before, or instead of, surgery. Two main types of radiotherapy are used to treat breast cancer: external radiotherapy and internal radiotherapy.

If part of the breast has been removed (lumpectomy or quadrantectomy), radiotherapy is usually given to the remaining breast tissue to reduce the risk of recurrence (19). The aim is to make sure that any remaining cancer cells are destroyed.

If all the lymph glands have been removed from under the arm radiotherapy to the armpit is not usually needed. If a few lymph glands have been removed and these contained cancer cells, or if no lymph glands have been removed, radiotherapy may be given to the armpit to treat the lymph glands.

Chemotherapy is the use of cytotoxic drugs to destroy cancer cells. Chemotherapy drugs are sometimes given as tablets or, more usually, intravenously. Chemotherapy is given as a course of treatment, which may last for less than one day or for a few days. This is followed by a rest period of a few weeks, which allows patients body to recover from any side effects of the treatment. The number of courses patient have will depend on the type of cancer she has and how well tumor is responding to the drugs. Side effects of chemotherapy includes: anemia, nausea and vomiting, hair loss, lowered resistance to infections, and diarrhea. Anthracyclines (doxorubicin, epirubicin), taxanes (paclitaxel, docetaxel), cyclophosphamide, 5-fluorouracil and methotrexate are the most commonly used chemotherapeutic agents used in breast cancer (20). Combinations of these agents have been used routinely for breast cancer treatment, most well known of which is, so called, "classical" CMF (cyclophosphamide, methotrexate and 5-fluorouracil) (21).

Hormonal therapies can slow or stop the proliferation of breast cancer cells by either altering the levels of estrogens which are naturally produced in the body, or preventing the hormones from being used by the cancer cells. There are many different types of hormonal therapy and they work in slightly different ways, so sometimes two different types of hormonal therapy may be given together. Hormonal therapy may also be given in combination with chemotherapy.

Most commonly hormonal therapies are: Anti-estrogen agents, Agents that reduce estrogen production, Progestogens, Pituitary down-regulators, and ovarian ablation).

Anti-estrogen agents work by preventing estrogen in the body from activating estrogen receptor and therefore inducing proliferation of tumor cells. Tamoxifen is the most commonly used hormonal therapy for breast cancer (22) and may be given in combination with other types of hormonal therapies known as aromatase inhibitors. The side effects which may be experienced include hot flushes and sweats, a tendency to put on weight, etc, but these side effects are usually mild. Rarely, it is possible for tamoxifen to cause an endometrial cancer Tamoxifen is commonly taken after surgery and for metastatic cancer, but if it is not effective in controlling the cancer some of the other types of hormonal therapy may be used. A drug called toremifene (Fareston) which works in a similar way to tamoxifen is occasionally used (22). Research and early tests suggest that it may carry less risk of endometrial cancer than tamoxifen, and it may be less likely to cause hot flushes and sweats. However, the long-term effects are not yet known. At the moment, toremifene is only given to postmenopausal women.

A group of agents called aromatase inhibitors work by blocking the production of estrogen in fatty tissues, in postmenopausal women. The commonly used aromatase inhibitors are anastrozole (Arimidex), letrozole (Femara), exemestane (Aromasin) and formestane (Lentaron). They generally do not cause many side effects, although they can cause hot flushes, feelings of nausea and joint pains. They are now sometimes used instead of tamoxifen as the first hormonal therapy (first-line treatment) in postmenopausal women with metastatic breast cancer (23).

Artificial progesterone derivatives (known as progestogens) are stronger than natural progesterone. Progestogens such as megestrol acetate (Megace) and medroxyprogesterone acetate (Farlutal, Provera) can be used if a hormonal therapy such as tamoxifen is no more effective. Progestogens generally cause few side effects (24).

Agents known as pituitary down-regulators, or LHRH analogues, reduce the production of estrogen-stimulating hormones by the brain, which results in a lowering of the level of estrogen in the body. This has the same effect as removing the ovaries or giving them radiotherapy, but is potentially reversible. As goserilin decreases the amount of estrogen circulating in the blood, it can also be an effective treatment for premenopausal women with metastatic breast cancer. Goserilin (Zoladex) only works for women with ER positive breast cancer. As goserelin brings on a temporary menopause, many of its side effects are similar to those of the menopause (25).

For premenopausal women, removing the ovaries (which reduces the level of estrogen in the body) can reduce the chance of the cancer coming back following surgery, or can slow the growth of cancer cells if they have already spread beyond the breast. The ovaries can be removed by a surgery, or stopped from working by giving a low dose of radiotherapy to the area (26). Unfortunately, removing the ovaries does bring on an early menopause which can be distressing, especially for a woman who was hoping to have children or complete her family.

Trastuzumab (Herceptin). ErbB2 (HER2/Neu) is amplified and overexpressed in 10-30% of breast cancers. Its amplification and overexpression have been associated with poor prognosis or response to anticancer therapies. Therapy based on a humanized monoclonal anti-ErbB2 antibody (trastuzumab/Herceptin™) has been beneficial in metastatic patients. Trastuzumab is the first monoclonal antibody with efficacy in breast cancer and the first oncogene-targeted therapy to yield a significant survival advantage in this disease. First-line trastuzumab in combination with chemotherapy resulted in a 25% improvement in overall survival compared with chemotherapy alone (27).

1.1.4. Biomarkers. Considerable efforts have been made to subdivide patient populations into groups that behave differently, so that therapy can be applied more efficiently. Already the early efforts of observation that outcomes were related to clinical cancer size and the presence or absence of pathologically involved lymph nodes, led to what is now commonly designated as "staging," which has now become highly codified within an internationally coordinated effort (17). These efforts only partially separate patients into subgroups with different biological behaviors. In the context of the development of molecular biology and biochemistry over the last decades laboratory and clinical scientists have studied a series of biomarkers. Hundreds of these putative markers have been reported, yet very few have actually achieved common clinical use. In part, this lack of progress is a consequence of the astonishing biological diversity of the disease. On the other hand, much of the perplexity and controversy in the field arises from poorly designed and analyzed clinical studies.

Currently used biomarkers could be grouped into risk assessment markers, prognostic markers and predictive markers.

Risk assessment markers. Roughly, 5-10% of breast cancers are caused by the inheritance of a germline mutation in a cancer predisposing gene. The most important of these genes are BRCA1 and BRCA2. Early data from highly selected families showed that women who carry either a BRCA1 or BRCA2 gene had an 80 to 85% lifetime risk of developing breast cancer (28). By testing for mutations in BRCA1 and 2 genes in high risk families, it is possible to identify individuals who are at increased risk of developing breast cancer. Identification of these subjects has the potential to result in early diagnosis and possibly prevention. However, genetic testing may also result in medical, psychological and other personal risks that must be addressed in the context of informed consent.

Prognostic markers can be defined as factors which correlate with patient outcome. If possible, these markers should be evaluated in the absence of adjuvant therapy. In breast cancer, prognostic factors are most useful in identifying patients whose outcome is so favorable that adjuvant systemic therapy is unnecessary. Prognostic indicators can also help in identifying patients whose prognosis is so poor with conventional approaches as to merit importance of more aggressive therapies. The traditional factors for assessing prognosis in breast cancer include tumor size, tumor grade and nodal status (29). Lymph node status is the most widely used, however, it has several disadvantages. Firstly, it requires major surgery. Major surgery does not result in enhanced survival compared to conservative surgery combined with radiotherapy. Another major disadvantage of nodal status for determining prognosis is that in the node-negative subgroup of patients, no reliable marker exists. With the development of mammographic screening, approximately two thirds of newly diagnosed breast cancer patients are node-negative. Approximately

70% of these patients are cured of breast cancer by surgery while the remaining 30% develop relapse within 10 years of diagnosis. Thus, new markers are urgently required which will differentiate the majority of node-negative breast cancer patients cured by surgery from the minority which develop relapse. Research in recent years has identified a large number of potential biologic prognostic markers for breast cancer (Table 1), of which, urokinase plasminogen activator (uPA) and its inhibitor PA1–1 are perhaps the most promising. More than 20 independent groups have reported that high levels of uPA predict unfavorable outcome in patients with breast cancer (30). In most of these studies the prognostic information supplied by uPA was independent of the traditional factors such as size, grade and nodal status. Furthermore, several different studies found uPA prognostic in node-negative patients. Paradoxically, high levels of PAI-1 have also been shown to predict poor outcome in breast cancer including the subgroup with node-negative disease. The prognostic impact of uPA/PA1–1 in node-negative patients was confirmed in a large prospective randomized trial, which provided the highest level of evidence (Level 1) for demonstrating clinical value for these cancer markers (31). Recently, pooled analysis of the EORTC-RBG datasets confirmed the strong and independent prognostic value of uPA and PAI-1 in primary breast cancer (32).

A *predictive marker* can be defined as a factor which predicts response or resistance to a specific therapy. The most widely used predictive marker in oncology is the estrogen receptor (ER) for selecting hormone responsive breast cancers (33). Although originally introduced to predict response to endocrine ablative therapy for patients with advanced breast cancer, the ER is now more widely used to select patients with early breast cancer likely to respond to the antiestrogen, tamoxifen. In a recent meta-analysis involving over 37,000 women, ER-positive patients were 7-times less likely to develop recurrent disease than ER-negative patients after at least 5 yr of adjuvant tamoxifen treatment (12). Assay of the progesterone receptor (PgR) may also help in selecting hormone-responsive breast cancers (33). Early work showed that patients with advanced breast cancer were more likely to respond to hormone therapy if their primary cancer expressed both ER and PR compared to those tumors containing ER but lacking PgR. Knowledge of PR status does not however, appear to enhance the predictive ability of ER in the adjuvant setting. Recent data suggests that ErbB 2 (also known as HER2 or *neu*) may also be a useful predictive marker in breast cancer. Preliminary findings suggest that overexpression of ErbB2 can select for resistance to hormone therapy, resistance to CMF (cyclophosphamide, methotrexate, 5-fluorouracil) adjuvant chemotherapy, sensitivity to doxorubicin-based adjuvant chemotherapy and response to the therapeutic antibody, Herceptin (34).

1.2. Signaling by Receptor Tyrosine Kinases and EGFR family.

1.2.1. Structure and function. Receptor tyrosine kinases (RTK) are type I membrane proteins, having their N- termini outside the cell and single transmembrane domain. N terminus starts with a signal peptide followed by an extracellular domain. Extracellular domain of RTKs is the most distinctive domain, which is composed of various recognizable sequence motifs and a pattern of Cys residues. Transmembrane domain is followed by a juxtamembrane region, which precedes the catalytic domain. The catalytic domain is about 250 residues long and is related to that of cytoplasmic PTKs. (35). The C terminal region varies from several up to 200 residues (Fig. 2).

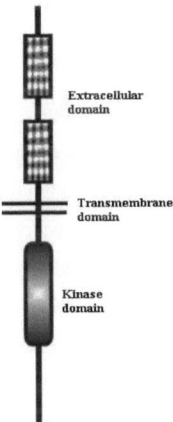

Figure 2. The schematic organization of receptor tyrosine kinase domains. N terminus is composed of an extracellular domain. Transmembrane domain is followed by a juxtamembrane region, which precedes the catalytic kinase domain. The length of C terminal region, following kinase domain, differs in various RTKs.

Within the catalytic domains of various RTKs, sequence similarity ranges from 32 to 95%. Alignment of kinase catalytic domains revealed there are 14 residues conserved in all protein kinases and several more which are found only in tyrosine kinases (36, 37). The structure of kinase domain is composed of two lobes. Mg^{2+}/ATP and the protein substrate are brought together in the cleft, which allows phophotransfer to be catalyzed. The N-terminal lobe is responsible for a Mg^{2+}/ATP binding. The responsible domain is the GXGXXG (21aa) K1030 (in human insulin receptor tyrosine kinase). The glycine fold holds the phosphate moieties of the nucleotide and the lysine residue, which is required for ATP binding. Further towards the C terminus another residue conserved in all protein kinases is E1047, which forms a salt bridge to K1030. In the other lobe, HRDLAARN (1130-1137) forms the catalytic loop. The Asp is believed to be the catalytic base. The aspartate of DGF (1150-1152) functions in the chelation of Mg^{2+}. E1179 and R1253 are thought to form ion bridges that stabilize the two lobes, and D1191 stabilizes the catalytic loop. (37, 38)

RTKs are known to function as receptors for growth (EGFR, PDGFR, FGFR), differentiation (MCF-R, NGFR) or metabolic (insulin receptor) factors. The functions of RTKs depend on several conditions, such as the cell specific expression of receptors as well as the availability of ligands and intracellular signaling molecules. Expression of almost all RTKs is restricted to specific cell types in the organism. This expression pattern depends on the character of the regulatory elements in the RTK gene promoter and enhancer. The function of RTK, expressed in particular cell is dictated by the ligands that bind the extracellular domain. It is also evident that the cell type in which the RTK is expressed affects the cellular response. The reason for this could be differences of substrates available, the strength of the signal and responses of cells to activation of the same pathway. On the other hand, different RTKs expressed in the same cell can evoke different responses through different signaling pathways.

1.2.2. Signaling. Signal transduction is initiated by ligand binding to the receptor. After ligand binding, the receptor dimerizes. Different ligands could use different approaches to induce active dimer. Some growth factors are dimers (VEGF, PDGF) and provide the simplest mechanism for receptor dimerisation (39), others are monomers and use more sophisticated mechanisms. FGF family ligands cooperate with accessory molecule heparin sulfate proteoglycan to activate FGF receptor dimers (40). Other ligands are thought to stabilize preexisting dimers. (41) Receptor dimerisation leads to trans-autophosphorylation. Dimerisation of extracellular domains leads to juxtaposition of the cytoplasmic tails, which leads to more efficient phosphorylation of tyrosines in the activation loop of the receptor. (42). Consequent to tyrosine phosphorylation, the activation loop adopts an "open" conformation that grants access to ATP and substrates, and enables phosphotransfer from MgATP to tyrosines on the receptor itself and on intracellular proteins involved in signal transduction. (38). The phosphorylated dimer recruits substrates that have an increased affinity for the phosphorylated tyrosine residues. Most tyrosine autophosphorylation sites are located in noncatalytic regions of the receptor molecule. These sites function as binding sites for SH2 (Src homology 2) or PTB (phosphotyrosine binding) domains of a number of signaling proteins (43) (Fig. 3). SH2 domains recognize distinct amino acid sequences determined by 1-6 residues C-terminal to the PY moiety (44), on the either hand PTB domains recognize PY within context of specific sequences 3-5 residues to its N terminus (45).

A big class of SH2 domain–containing proteins has intrinsic enzymatic activities such as PTK activity (Src kinases), phospholipase activity (PLCgamma), or Ras-GAP. Another class of proteins contains only SH2 and other modular domains. These adaptor proteins (Grb2, Nck, and Shc) apply their modular domains to mediate interactions of different proteins involved in signal transduction. For example, the adaptor protein Grb2 interacts with activated RTKs by its SH2 domain and recruits the guanine nucleotide releasing factor Sos close to its target protein Ras and therefore links the receptor to the Ras/MAPK pathway (46).

The binding of SH2 or PTB domain-containing proteins to phosphorylated RTKs can affect their activity in three ways: by membrane translocation, by the change of conformation or by tyrosine phosphorylation. Good example of activation by translocation to membrane is PI3K activation, while activation of Src family of PTKs is a classical example of activation by conformational change (see below). PLCgamma activation requires tyrosine phosphorylation as well as membrane translocation (see below), suggesting that three ways of activation are by no means mutually exclusive.

During the last decade, analyses of the different signaling cascades induced by RTKs let to the recognition of Ras/MAPK (47, 48), PI3K/Akt, PLCgamma/PKC (49, 50), and Src family PTKs (51) pathways as major downstream mediators of the RTK signaling. Several pieces of evidence suggest JAK/STAT pathway to play a very important role mediating RTK signaling as well (52, 53).

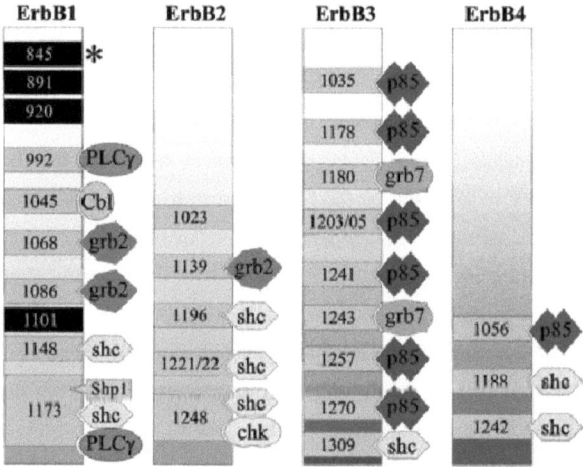

Figure 3. Authophosphorylation sites and binding of SH2 domain-containing signaling molecules to the ErbB family of RTKs (EMBO J. 2000 Jul 3;19(13):3159-67).

The Ras/MAPK pathway. All RTKs are known to stimulate the exchange of GTP for GDP on the monomeric G protein Ras, thus activating it. Biochemical studies in the cell lines and genetic studies in *Drosophila* have established that Ras is activated by the guanine nucleotide exchange factor, Sos. The adaptor protein Grb2 forms a complex with Sos, then Grb2/Sos complex is recruited to an activated RTK by Grb2 SH2 domain, thus translocating Sos to the plasma membrane where it is close to Ras and can catalyze exchange of GTP for GDP (46). Alternatively membrane recruitment of Sos can be also accomplished by binding of Grb2/Sos to Shc, another adaptor protein that forms a complex with many receptors through its PTB domain (45). Once activated, Ras interacts with several proteins, namely Raf. Activated Raf excites MAP-kinase-kinase (MAPKK, MEK) by phosphorylating a Ser residue in its activation loop. MAPKK then phosphorylates MAPK (ERK1/2) on T and Y residues at the activation-loop leading to its activation. Activated MAPK phosphorylates a variety of cytoplasmic substrates (MAPKAP) and, when translocated into the nucleus, it also phosphorylates transcription factors (54). MAP kinase family members have been found to regulate diverse biological functions by phophorylation of specific target molecules (such as transcription factors, other kinases, etc.) found in cell membrane, cytoplasm and nucleus, and thereby participate in the regulation of a variety of cellular processes including cell proliferation, differentiation, apoptosis and imunoresponses (55, 56).

The PI3K/Akt pathway. The class IA phospholipid kinase PI-3 kinase is activated by most RTKs. Like other SH2 domain–containing proteins, PI-3 kinase forms a complex with PY sites on activated receptors or with tyrosine phosphorylated adaptor proteins such as Shc. Activated PI-3 kinase phosphorylates PtdIns(4)P and PtdIns(4,5)P2 to generate the second messengers PtdIns(3,4)P2 and PtdIns(3,4,5)P3. The primary *in vivo* substrate is PtdIns(4,5)P2 (PIP2), which is converted to PtdIns(3,4,5)P3 (PIP3). The class IA PI3Ks consist of 2 subunits: regulatory – p85 and catalytic p100. p85 is an adaptor-like protein that has two SH2 domains and an inter-SH2 domain that binds constitutively to the p110 catalytic subunit.

The primary function of PI3K activation is the generation of PIP3, which functions as a second messenger to activate downstream tyrosine kinases Btk and Itk, the Ser/Thr kinases PDK1 and Akt (PKB) (57).

Signaling by the Akt kinase is described below.

The PLCgamma pathway. PLCgamma is immediately recruited by an activated RTK via the binding of its SH2 domains to PY sites of the receptor. When activated PLCgamma hydrolyzes its substrate PtdIns(4,5)P2 and forms two second messengers, diacylglycerol and Ins(1,4,5)P3. Ins(1,4,5)P3 stimulates the release of Ca^{2+} from intracellular stores. Ca2+ then binds to calmodulin, which in turn activates a family of calmodulin-dependent protein kinases (CamKs). Furthermore, both diacylglycerol and Ca2+ activate members of the protein kinase C (PKC) family. The second messengers generated by PtdIns (4,5)P2 hydrolysis stimulate a variety of intracellular such as proliferation, angiogenesis, cell motylity (58).

The Src protein tyrosine kinase family. Eight family members have been identified in mammals (59). At the N-terminal part they contain SH-2, and at the C-terminal part catalytic domain, followed by regulatory tyrosine phophorylation site (Y527 in c-src). When Y527 is phosphorylated, SH-2 domain of c-src itself binds to it, therefore folding kinase in inactive conformation (60). This conformation could be released by Y527 dephosphorylation or by binding of c-src SH-2 domain to phosphorylated tyrosines on active RTKs (61). SH-2 domain binds to phosphotyrosine sites on RTKs with higher affinity than to Y527 and thus allows substrates to access the catalytic domain. The activation of Src family kinases appears to be essential for mitogenic signal induced by many RTKs, such as PDGFR, NGFR and FGFR. However, the downstream substrates, involved in Src signaling, remain elusive. There is evidence, that src activates Ras. It is also known that p85 subunit of PI3K can interact with Src, Lyn, and Lck inducing PI3K activity (62, 63). Other substrates know to date include estrogen receptor alpha, p130Cas, ras-GAP, catenin p120, RACK1, etc. (64-68).

The JAK/STAT pathway. The binding of cytokines to their receptors leads to activation of JAK tyrosine kinases and following tyrosine phosphorylation of STATs. The SH2 domain of STAT binds to PY sites on other STAT implementing formation of STAT homodimers or heterodimers. The dimeric STATs move to the nucleus and function as a transcription factor (69). There is good indication that JAK/STAT signaling plays a role in RTK signal transduction. PDGF, EGF, ErbB4 or IGF stimulation leads to rapid tyrosine phosphorylation and migration of STATs, namely STAT1, STAT3 and STAT5, to the nucleus. (52, 70-72) There were several different mechanisms of STAT activation by RTKs described, such as direct phosphorylation or even binding of STATs to RTKs (73-75), activation through Src tyrosine kinase (76-78) and "classical" activation through JAK family tyrosine kinases (53, 78).

1.2.3. ErbB family of RTKs. The type I receptor tyrosine kinases or the ErbB family consists of four members who are named for their homology to the v-erbB oncogene: ErbB1 (EGFR, HER1), ErbB2 (HER2/Neu), ErbB3 (HER3) and ErbB4 (HER-4) (Fig. 4). They have molecular weight of 170-185 kDa and share two structural aspects by which they can be distinguished form the other receptor tyrosine kinases: two cysteine rich clusters in the extracellular region and an uninterrupted tyrosine kinase domain in the cytoplasmic part (79, 80).

The EGFR was the first member of the ErbB family and the first of RTKs to be cloned and sequenced. It was also the first receptor for which ligand-dependent activation was demonstrated (81). The EGFR gene is located on chromosome 7p13-q22 and codes for a protein of 1210 amino acids which weighs, when glycosylated, 170 kDa. Ligands which bind to the EGFR represent a family of growth factors, called the EGF family. This family consists of such members as EGF, TGFα, HB-EGF, amphiregulin betacellulin and epiregulin (82-84). Betacellulin, HB-EGF and epiregulin also bind to ErbB4 (85-87). The EGFR is the only family member, which is internalized in coated pits upon ligand binding, receptor dimerisation and activation. Internalization is followed by lyzosomal degradation (88), and partial inactivation through phosphorylation of serine and threonine residues within the intracellular domain (89, 90). The EGFR is expressed in a variety of normal tissues, including normal breast tissue. Its importance has been emphasized by the lethal knock-out mice. A large number of deletion variants of EGFR mRNA have been observed in various types of cancer, including breast cancer (91) and ovarian cancer (92). These deletions are the result of genomic rearrangements, resulting in alternative splicing of the mRNA. They are found both in part of mRNA which encodes the extracellular as well as intracellular regions of the EGFR (93) giving rise to truncated and often constitutively active receptors (94, 95).

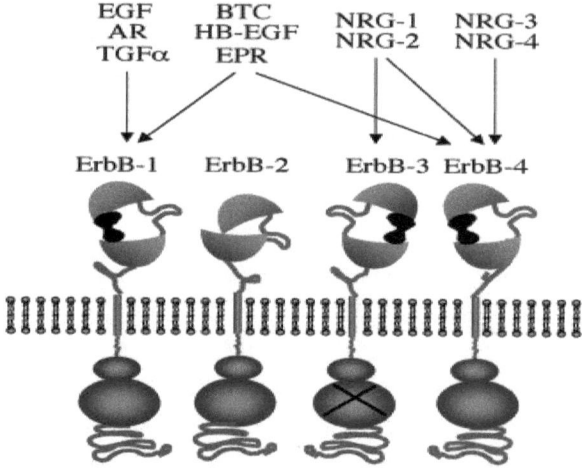

Figure 4. ErbB family members (Int J Radiat Oncol Biol Phys. 2004 Mar 1;58(3):903-13).

The ErbB2 gene is located on chromosome 17q21 and encodes a protein of 1255 amino acids which weighs, when glycosylated, 185 kDa. The human ErbB2 was cloned by homology screening with v-ErbB (96) and has the highest homology to the EGFR among ErbB family members. It is mostly related to EGFR in its kinase domain (82%) and mostly distinct in the C-terminus, which contains most of the autophosphorylation sites. ErbB2 is the only orphan receptor of the ErbB family, since no ligand binding it has been found up to date. Activation of ErbB2 is therefore highly dependent on the expression of other family members, to which it is recruited as a preferred heterodimeric partner (97). Alternatively, overexpression and/or mutation of ErbB2 are thought to lead to spontaneous dimerisation and the stabilization of the receptor dimmers in a ligand-independent manner (98-100). Like the EGFR, partial inactivation of the ErbB2 tyrosine kinase is mediated by PKC through phosphorylation of serine and threonine residues within the intracellular part of the receptor (101). ErbB2 is amplified and/or overexpressed in a number of human cancers including gastric, esophageal, salivary, colon, bladder and lung cancers (102, 103). ErbB2 overexpression correlates with tumor progression and aggressiveness, poor prognosis and an elevated metastatic potential.

ErbB3 and ErbB4 were both found by homology screening. The genes of ErbB3 and ErbB4 are located on chromosomes 12q13 and 2q33 respectively and code for proteins of 1342 and 1308, both of which weigh, when glycosylated, 180 kDa (104, 105). The ligands for ErbB3 and ErbB4, are the various isoforms of the neuregulins (NRG).

The rate of catalysis of the ErbB3 tyrosine kinase is only 1% of that of the other receptors in the family (106). There are four amino acid changes in the kinase domain of ErbB3. These four amino acids differ from the sequences of all known protein kinases. The change of Asn to Asp is particularly important, since it is responsible for the loss of ErbB3 kinase activity. The same amino acid substitution in other tyrosine kinases resulted in the loss of function as well (107). Interesting feature of ErbB3 is the presence of seven YXXM repeats in the carboxy terminus. These authophosphorylation sites serve as docking sites for PI3K. Such motifs are missing in the other family members and confer a specific signaling ability on ErbB3 (108). The expression of ErbB3 is in general different form that of EGFR and ErbB2, since it is frequently expressed in differentiated cells. It is particularly important in the peripheral nervous system and in neuromuscular synapse formation (109-111). It has been found both overexpressed and underexpressed in DCIS (112) and in breast cancers (106). Prognostic value of ErbB3 is controversial.

ErbB4 is a unique member of ErbB family: it exists in two isoforms. Sequencing of full-length human ErbB4 from either a human MDA-MB-453 breast cancer cell line (104) or from human fetal brain tissue (113) revealed the presence of two isoforms, JM-a and JM-b, that differ by insertion of either 23 or 13 alternative amino acids in the juxtamembrane region. The two isoforms differ in their expression pattern: both are expressed in neural tissues, whereas kidney expresses only JM-a and heart only JM-b (114). The difference in the juxtamembrane region did not alter the extent of activation by hereguluins. Nevertheless, a functional difference was observed upon phorbol ester treatment. Treatment of JM-a transfected cells, but not JM-b transfected cells resulted in a loss of HRGβ1 binding and reduction in total cell-associated ErbB2 protein levels. JM-a may thus represent a cleavable receptor form. ErbB4 was found to be downregulated by phorbol esters by activating a selective proteolytic mechanism. Proteolytic cleavage produces an 80 kDa cytoplasmic domain fragment and a 120 kDa ectodomain fragment. Cytoplasmic fragment of ErbB4 is dephosphorylated when cleaved, therefore the 80 kDa fragment is not an active tyrosine kinase. The function of the proteolytic cleavage id not known, but is thought to direct the cytoplasmic domain to proteosome (115). ErbB4 is expressed in several adult tissues including heart, kidney, brain and skeletal muscle (104). ErbB4 knock-out mice die at embryonic day 10-11 and have severe cardiac and neural defects (116). Overexpression of ErbB4 has been observed in 10-20% adenocarcinomas of the breast, colon, ovary, prostate and endometrium, whereas underexpression were found in 40-80% of the malignancies, reaching 100% in squamous cell carcinomas of the head and neck (117).

1.2.4. Crosstalk between ErbB family and estrogen receptor signaling. Numerous in vitro studies demonstrate that hormone-responsive breast cancer cells with upregulated ErbB2 display resistance to tamoxifen. On the other hand, reports differ on the effect of ErbB2 on estrogen dependence. It has been reported that ErbB2-overexpressing MCF-7 cells remained estrogen dependent but became tamoxifen resistant (251). Others report, that ErbB2 overexpression reduces dependence on estrogen (252, 253). In vivo work in general supports the idea that breast cancer cells containing both ER and ErbB2 are sensitive to estrogen withdrawal, and resistant to tamoxifen, but over time tumor xenografts develop an estrogen-independency. Altogether, these data propose a role for ErbB2 in the survival and proliferation of ER positive breast cancers under low estrogen environment. A role for EGFR in endocrine therapy resistance has also been established in both preclinical and clinical studies (254, 255). Multiple experiments have demonstrated that inhibition of ErbB1 and/or ErbB2 with either trastuzumab or a selective tyrosine kinase inhibitor can reverse tamoxifen resistance (256-259). Identical effect can be achieved through inhibition of the downstream signaling molecules Akt (260) and MAPK (261). Moreover, in some cell culture models, long-term exposure to estrogen withdrawal induces ErbB2 upregulation and produce ErbB2-dependent resistance (262). Preliminary data support the idea that such event may take place throughout the development of human breast cancer (263). However, the high expression of ErbB2 alone is not a major factor in determining drug resistance in breast cancer cells. Coexpression of either EGFR or ErbB3 with ErbB2 significantly enhanced drug resistance (264).

The biochemical details of the cross-talk between the ErbB2 and ER pathways are unclear. The effect of ErbB2 may be to optimize as well as enhance DNA binding, interaction with coactivators and transcriptional activity of the ER. On the other hand, ErbB2 activates the Ras/MAPK and PI3K/Akt pathways, which may be involved in post-translational modification of ER (265-267). It is also likely that a combination of post-translational modifications as well as alterations in the assembly of multi-component transcription complexes may occur.

Since both ER and ErbB2 positive tumors are relatively uncommon, many studies have too few cases to make strong conclusions regarding the clinical behavior associated with this biomarker profile. However, results from studies in the advanced disease setting do suggest that patients whose tumors overexpress ErbB2 have poorer outcomes after the endocrine therapy than patients whose tumors express lower levels of ErbB2 (268). Moreover, studies in the adjuvant and neoadjuvant settings support a role for ErbB2 and, to a lesser extent, ErbB1 in tamoxifen resistance. Several studies of tamoxifen as adjuvant therapy suggest that patients with ErbB2 tumors receiving tamoxifen may even have worse outcome than patients receiving placebo (269). In contrast, other studies show that patients with both ER and ErbB2 positive tumors can have pretty good outcome with endocrine therapy, especially when estrogen withdrawal is a component of the treatment strategy (270). Moreover, in a neoadjuvant study for postmenopausal women with ER positive locally advanced disease, patients with tumors typed as ER, ErbB1, and/or ErbB2 positive responded well to letrozole but poorly to tamoxifen (271). These data imply that estrogen withdrawal might be an important adjuvant strategy for tumors that are ErbB2 (possibly ErbB1 as well) and ER positive.

1.3. Shc adaptor proteins.

Shc was identified by low-stringency hybridizations to human cDNA libraries, using an SH2-coding sequence as a probe (118). The originally cloned *shc* transcript (ShcA) displayed two in-frame ATGs and was shown to encode two polypeptides: the ubiquitously expressed p52ShcA and p46ShcA proteins. These two isoforms share an amino-terminal SH2 domain, followed by a CH1 domain, and a carboxy-terminal phosphotyrosine-binding domain (PTB). A third ShcA isoform, p66ShcA, was later characterized (119) (Fig 5).

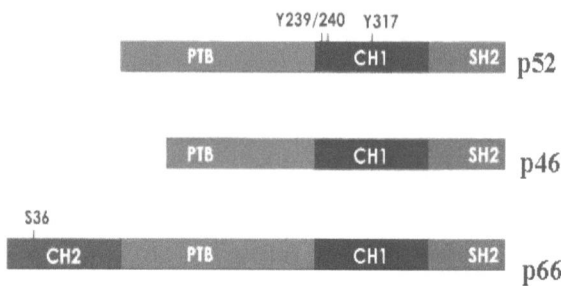

Figure 5. The schematic organization of ShcA isoforms. The PTB and SH2 domains of ShcA bind to phosphotyrosine containing sequences. Within the CH domain, three tyrosine-phosphorylation sites have been identified. The p46Shc lacks the first 46 amino acids within the PTB domain. The p66Shc possesses an additional CH2 domain that contains a serine phosphorylation site (Oncogene. 2001 Oct 1;20(44):6322-30).

Biochemical and genetic data prove a general role of p52/46ShcA proteins in the transduction of signals from tyrosine kinases (TKs) to Ras (120-121). p52/46ShcA proteins are effectively tyrosine-phosphorylated by all TKs tested to date. Three major ShcA phosphorylation sites have been identified, which are all located within the CH1 region (Y239–240 and Y317). The role of phosphorylated p52/46ShcA proteins is to link together RTKs and the SH2-containing Grb2 adaptor protein (122-124). The PTB and/or SH2 domains of ShcA bind to tyrosine-phosphorylated receptors whereas the SH2 of Grb2 binds the tyrosine-phosphorylated ShcA. Grb2, in turn, is constitutively bound to Sos, a ubiquitously expressed Ras guanine nucleotide exchange factor. Recruitment of the Grb2/Sos complex, by p52/46ShcA, results in the membrane relocalization of Sos and subsequent Ras activation. The centrality of p52/p46ShcA in Ras activation is, however, challenged by a several of observations. First, the Grb2/Sos complex can be recruited directly to activated receptors with ensuing Ras activation. Second, in cells, derived from Shc knockout mouse embryos, Ras activation seems to occur normally (125). It has been suggested that p52/p46ShcA might serve as 'amplifiers' of RTK signaling, in the pathway leading to Ras activation at low concentration of growth factors. However, the precise role of ShcA in the activation of Ras is still obscure.

Alternatively, ShcA proteins might serve other and evidently unrelated functions. Homozygous mutation of p66ShcA in mice was shown to cause increased resistance to oxidative-stress-induced apoptosis and life-span extension (126). p66ShcA is a third isoform encoded by the human and mouse *shc* loci through alternative splicing (119). It contains the entire p52/46ShcA sequence and an additional domain, similar to CH1 domain, named CH2. Regardless of its tyrosine-phosphorylation by active RTKs, p66ShcA is not involved and even seems to inhibit Ras activation (127). Instead, it is involved in pathways activated by environmental stresses, as shown by its serine-phosphorylation within the CH2 domain.

ShcA proteins were shown to be independent prognostic markers for primary breast cancer (272). IHC staining intensities demonstrated that increased amounts of PY ShcA and decreased protein expression levels of p66ShcA protein correlated with disease recurrence. The ratio of PY ShcA to p66ShcA was 2-fold higher in primary tumors of patients who subsequently relapsed.

Growing complication in the functions of Shc is further projected by the recent identification of two human Shc homologues —ShcB and ShcC — which share the same PTB–CH1–SH2 modular organization (128, 129). Two isoforms of ShcC have been identified — p64ShcC and p52ShcC — which are encoded by the same transcript by alternative usage of in-frame ATGs. Preliminary evidence suggests also that ShcB and ShcC are RTK substrates, which bind activated receptors through their PTB/SH2 domains. Unlike ubiquitous ShcA, ShcB and ShcC are specifically expressed in the brain.

1.4. Akt signaling.

1.4.1. Structure. Akt, also termed Protein kinase B (PKB) is a serine/threonine kinase, which belongs to the 'AGC' superfamily of protein kinases. Akt, as other AGC kinases, is regulated by upstream second messengers as well as other enzymes. For Akt, this activating process involves multiple inputs that strictly control the place, length and power of response. In mammals, there are three isoforms of Akt: Akt1, Akt2 and Akt3 (PKBα, PKBβ, PKBγ;) (Fig.6). All three isoforms share a high degree of amino acid identity and are composed of three functionally different regions: an N-terminal pleckstrin homology (PH) domain, a central catalytic (kinase) domain, and a C-terminal hydrophobic motif (HM). This general structure is conserved across species including *Drosophila melanogaster* and *Caenorhabditis elegans*, suggesting that regulation of Akt appeared early during the evolution.

The catalytic domain of Akt is structurally related to other protein kinases of the AGC family (130).There are two important regulatory domains, which control the activity and specificity of the protein kinase domain. The N-terminal PH domain is common to various signaling proteins and provides a lipid binding element to direct Akt to PI3K-generated phosphoinositides PI(3,4,5)P3 and PI(3,4)P2. Therefore, growth factors that increase PI3K activity provide a plasma membrane recruitment mechanism for Akt.

The crystal structure of the PH domain of PKB bound with the inositol head group of PI(3,4,5)P3 has been solved (131). Interestingly, this structure revealed differences in the binding between inositol(1,3,4,5)P4 and other PI(3,4,5)P3 binding PH domains, such as BTK. The most significant of these differences is that the D5 phosphate of inositol(1,3,4,5)P4 does not physically interact with the Akt PH domain, in agreement with earlier studies demonstrating that PI(3,4,5)P3 and PI(3,4)P2 bind to Akt with equal affinities. This has important implications regarding the upstream phosphatase regulators of PI3K-generated lipids, PTEN (132, 133) and SHIP (134). Since PTEN catalyzes the dephosphorylation of PI(3,4,5)P3 and PI(3,4)P2 at the D3 position, the actions of this phosphatase would reduce the entire pool of lipids capable of binding with Akt.

Another important regulatory domain of Akt is a C-terminal HM, present in many other AGC kinases (135). The HM provides a docking site for the upstream activating kinase, 3-phosphoinositide-dependent kinase-1 (PDK1) (136-138). Interrupting this docking interaction severely attenuates phosphorylation of the activation loop (T-loop) of AGC kinases. The HM also serves as an allosteric regulator of catalytic activity (130, 137, 139). The HM provides stability to the catalytic core by association with hydrophobic and phosphate binding pockets created by a cleft formed at the junction of the αB-helix, αC-helix and β5-sheet in the N-lobe of the kinase domain. Stabilization of the N-lobe of the kinase domain by binding of the HM, increases the phosphotransfer rate up to 10-fold. Mutations of key phenylalanine residues within the HM compromise catalytic activity of Akt (140).

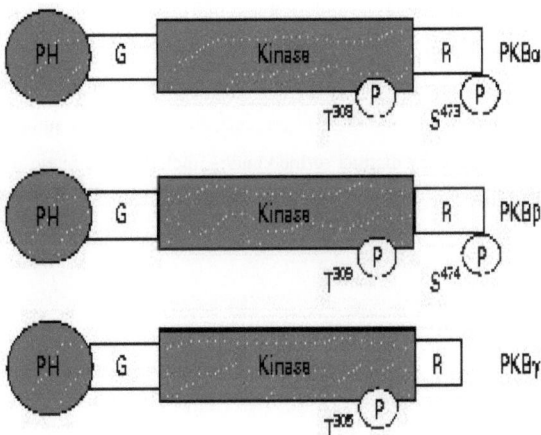

Figure 6. The schematic organization of Akt isoforms. All the Akt isoforms possess the kinase domain in the central region of the molecule. The PH (pleckstrin homology) domains act as phosphoinositide-binding modules. The hydrophobic regulatory (R) motif is located at the carboxyl-terminal adjacent to the kinase domain.

1.4.2. Phosphorylation. PKB is activated by, and dependent upon, multisite phosphorylation. The main site of phosphorylation is within the activation T-loop at Thr308 (for Akt1). Phosphorylation of Akt on Thr308, causes a change in conformation allowing substrate binding and greatly elevated rate of catalysis. There is very low phosphorylation of the activation loop in resting cells and rapid increase in phosphorylation upon agonist stimulation. The phosphorylation of Thr308 strictly regulates the activation of Akt, and its mutation to alanine impairs the kinase activity (141). It has been established that Thr308 is phosphorylated by the PDK1 (142, 143). PDK1 phosphorylates Akt in vitro, and overexpression of PDK1 in cells also leads to elevated Thr308 phosphorylation in the absence of natural agonists. PDK1 contains a C-terminal PH domain, and the rate of Akt phosphorylation by PDK1 is significantly increased in vitro by the addition of PI(3,4,5)P3 or PI(3,4)P2, which recruits both to a plasma membrane (142). In cells in which PDK1 has been disrupted, Akt is unresponsive to mitogenic stimulation as a result of a loss of Thr308 phosphorylation (144, 145).

An additional layer of regulation is provided by HM phosphorylation, namely Ser473 phosphorylation (for Akt1). The mechanism of Ser473 phosphorylation is not completely understood, and there is evidence suggesting both autophosphorylation (146) and phosphorylation by distinct serine kinases, including the integrin-linked kinase (ILK) (147, 148).

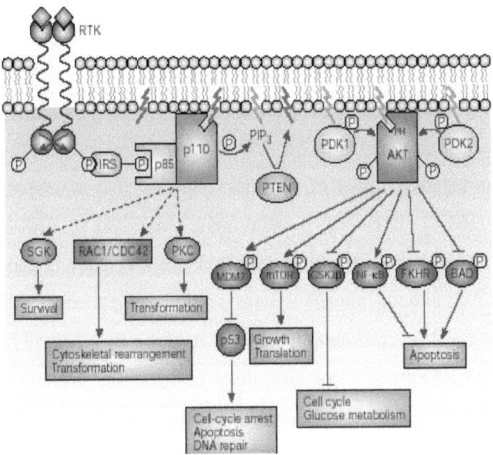

Figure 7. Activation of class IA phosphatidylinositol 3-kinases (PI3Ks) occurs through stimulation of receptor tyrosine kinases (RTKs) and the concomitant assembly of receptor–PI3K complexes. These complexes localize at the membrane where the p110 subunit of PI3K catalyses the conversion of PtdIns(4,5)P2 (PIP2) to PtdIns(3,4,5)P3 (PIP3). PIP3 serves as a second messenger that helps to activate AKT. Through phosphorylation, activated AKT mediates the activation and inhibition of several targets, resulting in cellular growth, survival and proliferation through various mechanisms (Nat Rev Cancer. 2002 Jul;2(7):489-501).

Several groups have investigated the possible role of tyrosine phosphorylation in Akt regulation (149-151). One report indicated that two tyrosine residues located within the catalytic domain of PKB, Y315 and Y326, are phosphorylated following receptor activation and are required for activity, since mutation to phenylalanine abolished kinase activity (149). A second report identified Y474, as a possible site of phosphorylation in response to insulin and pervanadate (151). Mutation of this residue to phenylalanine reduced Thr308 phosphorylation and reduced Akt activation by about 50%. A very interesting point is that the authors show that S473 and Y474 phosphorylation are mutually exclusive: phosphoamino acid analysis and N-terminal sequencing of the tryptic peptide containing S473 and Y474 show that neither are phosphorylated together.

1.4.3. Functions. The main biological functions of Akt activation can be grouped into three categories — survival, proliferation and cell growth.

Survival. Apoptosis is a normal cellular function that controls cell numbers by eliminating 'excessive' cells. Cancer cells have devised several mechanisms to inhibit apoptosis and prolong their survival. Akt functions in an anti-apoptotic pathway, since dominant-negative alleles of Akt block survival that is mediated by insulin-like growth factor 1 (IGF1) (152), and constitutively active Akt rescues PTEN-mediated apoptosis (153). The mechanism by which Akt prevents cell death is expected to be multifactorial, because Akt directly phosphorylates several components of the apoptotic system. For example, BAD is a pro-apoptotic member of the BCL2 family of proteins that promotes apoptosis by forming a non-functional hetero-dimer with the survival factor BCL-X_L. Phosphorylation of BAD by Akt prevents this interaction (154), restoring BCL-X_L's anti-apoptotic function. Likewise, Akt inhibits the catalytic activity of a pro-death protease, caspase-9, through phosphorylation (155). Lastly, phosphorylation of members of the Forkhead family of transcription factors by Akt prevents its nuclear translocation and activation of gene targets (156), which include several pro-apoptotic proteins. AKT can also influence cell survival by means of indirect effects on two central regulators of cell death — nuclear factor of κB (NF-κB) (157, 158) and p53 (159, 160).

Proliferation. The cell cycle is regulated by the coordinated action of cyclin–cyclin-dependent kinase (CDK) complexes and CDK inhibitors (CKIs). Cyclin D1 levels, which are important in the G1/S phase transition, are regulated at the transcriptional, post-transcriptional and post-translational level by distinct mechanisms. Akt has an important role in preventing cyclin D1 degradation by regulating the activity of the cyclin D1 kinase glycogen synthase kinase-3β (GSK3β). After phosphorylation by GSK3β, cyclin D1 is targeted for degradation by the proteosome. Akt directly phosphorylates GSK3β and blocks its kinase activity, thereby allowing cyclin D1 to accumulate (161). Akt can also negatively influence the expression of cylin kinase inhibitors, such as p27 (KIP1) and p21 (CIP1 or WAF1) (162). Akt can also modulate p21 and p27 activity by affecting its phosphorylation either directly or through intermediate kinases. The functional importance of connections between Akt and the cell-cycle are supported by experiments showing that the blockade of PI3K or Akt activity leads to cell-cycle arrest (163-164).

Cell growth. In addition to its role in proliferation, there is increasing evidence that Akt also affects cell growth. A central regulator of cell growth is mTOR (the mammalian target of rapamycin), a serine/threonine kinase that serves as a molecular sensor that regulates protein synthesis on the basis of the availability of nutrients. mTOR regulates biogenesis by activating p70 S6 kinase, which enhances the translation of mRNAs that have 5' polypyrimidine tracts, and by inhibiting — a translational repressor of mRNAs that bears a 5' CAP structure. mTOR is a direct target of Akt (165), however, it is still unclear how or whether phosphorylation of mTOR by Akt is a mechanism for activation. Pharmacological studies with the mTOR inhibitor rapamycin indicate that the Akt pathway regulates cell growth through mTOR. Yet, the PI3K–AKT pathway is unlikely to be the only stimulus that leads to mTOR activation in cancer cells.

2. EXPERIMENTAL PROCEDURES

2.1. Materials.

Separating Gels, in 0.375 M Tris, pH 8.8

	7%	10%	12%	15%
distilled H2O	5.15 ml	4.15 ml	3.45 ml	2.45 m
1.5 M Tris-HCl, pH 8.8	2.5 ml	2.5 ml	2.5 ml	2.5 ml
Acrylamide/Bis-acrylamide (30%/0.8% w/v)	2.3 ml	3.3 ml	4.0 ml	5.0 ml
25% (w/v) ammonium persulfate	0.05ml	0.05ml	0.05ml	0.05ml
TEMED	0.005 ml	0.005 ml	0.005 ml	0.005 ml
Total:	10ml	10ml	10ml	10ml

Stacking Gels, 4.0% gel, 0.125 M Tris, pH 6.8

distilled H2O	3.1 ml
1.5 M Tris-HCl, pH 8.8	1.25 ml
Acrylamide/Bis-acrylamide (30%/0.8% w/v)	0.67 ml
25% (w/v) ammonium persulfate	0.025ml
TEMED	0.005 ml
Total:	5.05 ml

Resolving gel buffer: 100 ml
0.4 g SDS (add last)
18.2 g Trizma base (= 1.5M)
Adjust pH to 8.8 with concentrated HCl

Stacking gel buffer: 100 ml
0.4 g SDS (add last)
6.05 g Trizma base (= 0.5 M)
Adjust pH to 6.8

10x Running buffer: 1 L
30.3 g Trizma base (= 0.25 M)
144 g Glycine (= 1.92 M)
10 g SDS (= 1%)--add last
Do not adjust the pH!!

Blotto: 0.5 L
2.5% Nonfat dry milk powder
Make up in PBS.
Then add 0.05% Tween 20.
Keep at 4°C to prevent bacterial contamination.

10x Transfer buffer: 1 L
30.3 g Trizma base (= 0.25 M)
144 g Glycine (= 1.92 M)
pH should be 8.3; do not adjust

To make 2 L of 1x Transfer buffer:
400 ml Methanol
200 ml 10x Transfer buffer
1400 ml water

Stripping buffer: 0.5 L
62.5 mM Tris-HCl pH 6.8
2% (w/v) SDS
100 mM b-mercaptoethanol

2.2. Cell culture. SKBr3 and MCF-7 breast cancer cell lines obtained from American Type Culture Collection (Manassas, VA), were cultured in Improved Minimal Essential medium with zinc option (IMEM-ZO) supplemented with 5% fetal bovine serum and L-glutamine at 37°C in a 5% CO_2 incubator.

For use as an immunoassay standard, subconfluent (70-80%) SKBr3 or MCF-7 cells were serum-starved for 48 h in serum free medium, treated with Na_3VO_4 or NaF for 1 hr., then with 10% FBS for 10 min. Cells were lysed in EB lysis buffer as described below.

For lysis, cells were washed 3X with ice-cold PBS. 0.5-1ml of EB buffer (0.5 M NaCl, 10 mM EDTA, pH 8, 1% Triton X100, 20 mM Tris-Cl, pH 7.0, 20mM NaF, 20mM glycerophosphate, 2 mM sodium pervanadate, 1 mM PMSF, phosphatase inhibitor cocktail (Roche)) was added to the cells and cells lysed for 5 min, on ice. Cells were then scraped using "rubber policeman" and lysate transferred to eppendorf tube and spun at 20 000g for 5 min. at 4°C. Supernatant was transferred to new eppendorf tube and stored at -80°C.

2.3. Tumor extract preparation for immunoassays. Tissue homogenates were prepared in accordance with standard procedures for tumor marker EIA measurement. The frozen tissues were pulverized in liquid nitrogen using a Micro-Dismembrator U (B. Braun Melsungen AG, Melsungen, Germany). The powder was homogenized with a tissue homogenizer (Ultra-Turrax; Janke & Kunkel, IKA-Werke, Staufen, Germany) for 20 seconds in three volumes of ice-cold extraction buffer containing Tris 10 mmol/L, EDTA 1.5 mmol/L, 10% glycerol, disodiummolybdate 5 mmol/L, and monothioglycerol 1 mmol/L. The homogenate was centrifuged for 3 minutes at 4°C, and the supernatant was recentrifuged in an ultracentrifuge (Beckman Instruments, Fullerton, CA) at 100,000g for 40 minutes at 4°C.

2.4. Measurement of ER and PgR levels. The cytosols, resulting from the above described procedure were used for measurement of the hormone receptors (ER, PgR). ER and PgR concentrations were measured from tumor cytosolic extracts by commercial quantitative ER and PgR EIA kits (Abbott Laboratories, Abbott Park, IL) using a Quantum II photometer.

2.5. Immunoassay of total ErbB2 receptor levels. ErbB2 receptor levels were determined on the particulate membrane fractions of tumor extracts using a commercial monoclonal antibody EIA kit. The cryopreserved pellet fraction obtained from the high-speed tumor extract (above) was washed once with cold magnesium chloride solution (0.5 mol/l) and three times with cold Tris hydrochloride buffer (10 mmol/L, pH 7.4) containing EDTA 1 mmol/l. Washed pellets were suspended on ice in 5 volumes of a buffer containing HEPES 20 mmol/l (pH 7.4), 10% glycerol, leupeptin 10 mg/l, and aprotinin 2 mg/l. The pellet suspension was then further diluted with HEPES-glycerol-aprotinin-leupeptin buffer containing 0.1% of Triton X-100. Protein concentrations of these diluted extracts were determined using the Pierce BCA (bicinchoninic acid) Protein Assay (Pierce, Rockford, IL). Quantitative assessment of ErbB2 protein levels was carried out with immunoassay kit (Oncogene Science kit; Oncogene Science, Cambridge, MA), according to the manufacturer's instructions. Color intensity was measured on a COBAS EIA spectrophotometer (Hoffmann-La Roche Ltd, Basel, Switzerland). Internal quality control was performed during each run using the controls provided with each kit.

2.6. Immunoassay of P-Y1248 ErbB2 levels. Black ninety-six–well microtiter plates (Nunc Black MaxiSorp Surface; Nalgen Nunc International, Rochester, NY) were coated with antihuman activated *neu*/c-*ErbB2* antibody no. 06-229 (lot no.15916; Upstate Biotechnology, Lake Placid, NY) at a concentration of 4 mg/mL of coating buffer (PBS with 0.6 mM EDTA) in a volume of 100 µl/well and kept at 4°C overnight. This rabbit polyclonal antibody was raised against a phosphorylated polypeptide containing the C-terminal autophosphorylation site of ErbB2, virtually identical to the phosphorylated epitope used to produce the monoclonal antibody clone PN2A that recognizes only PY1248 ErbB2 used in several previous studies (25-29). For determination of phosphorylated ErbB2 tumor extracts were prepared as described above in the presence of Na_3VO_4. Before sample applications, the coated microtiters were washed five times with 200 µl/well of washing buffer (25mM HEPES, pH 7.4, 300 mM NaCl, 0.05% Tween-20) and then blocked for 2 hours at room temperature with 250 µl of blocking buffer (25mM HEPES, pH 7.4, 300 mM NaCl, 0.05% Tween-20, 3% TopBlock, Juro AG, Switzerland). The blocked wells were washed five times with 300 µl of blocking buffer and then 100 µl of the diluted tumor membrane extracts or reference material was added to the wells and incubated overnight at 4°C. As a reference for each assay, a s cell extract of SKBr3 cells, stimulated as described above, was used. For use in ELISAs, the SKBr3 cell membrane extract was sequentially diluted with sample dilution buffer at ratios of 1X, 0.75X, 0.5X, 0.25X, 0.125X, and 0.025X, and then 100 µl aliquots were incubated on each microtiter plate together with the tumor tissue extracts and blanks (containing only dilution buffer). After incubation of the samples and references, the wells were washed five times with 300 µl washing buffer at room temperature to eliminate unbound particles. Biotinylated detection antibody (detecting antibody from HER-2/neu Microtiter ELISA kit, OncogeneScience) was added to the wells and incubated for 2 hrs at room temperature. Complex was detected with horseradish peroxidase (HRP)-conjugated streptavidin, diluted in conjugate diluent for 1 hr at room temperature. HRP activity was detected with SuperSignal WestPico substrate (Pierce, Rockford, IL) in glow luminometer. The response data of the series of diluted reference material were fitted and the curve was used for quantification of the tumor extracts. The value of the undiluted SKBr3 extract was denominated 100 U/ml.

2.7. Immunoassays of pan-Y, S and T phosphorylation of ErbB2. Black ninety-six–well microtiter plates (Nunc Black MaxiSorp Surface; Nalgen Nunc International, Rochester, NY) were coated with monoclonal Anti-c-ErbB2 no. E2777 (clone HER2-96, Sigma,) at a concentration of 4 mg/mL of coating buffer in a volume of 100 µl/well and kept at 4°C overnight. For determination of phosphorylated ErbB2 tumor extracts were prepared as described above in the presence of Na_3VO_4. Before sample applications, the coated microtiters were washed five times with 200 µl/well of washing buffer and then blocked for 2 hours at room temperature with 250 µl of blocking buffer. The blocked wells were washed five times with 300 µl of blocking buffer and then 100 µl of the diluted tumor membrane extracts or reference material was added to the wells and incubated overnight at 4°C. As a reference for each assay, a s cell extract of SKBr3 cells, stimulated as described above, was used. For use in ELISAs, the SKBr3 cell membrane extract was sequentially diluted with sample dilution buffer at ratios of 1X, 0.75X, 0.5X, 0.25X, 0.125X, and 0.025X, and then 100 µl aliquots were incubated on each microtiter plate together with the tumor tissue extracts and blanks (containing only dilution buffer). After incubation of the samples and references, the wells were washed five times with 300 µl washing buffer at room temperature to eliminate unbound particles. Biotinylated PhosphoSerine Antibody Q5, Cat. No. 37430 or PhosphoThreonine Antibody Q7, Cat. No. 37420 (Qiagen, Chatsworth, CA) or alkaline phosphatase (AP) conjugated PhosphoTyrosine antibody PY20, Cat. No. 03-7722 (Zymed Laboratories Inc., South San Francisco, CA) was added to the wells and incubated for 2 hrs at room temperature. Complex was detected with AP-conjugated streptavidin, diluted in conjugate diluent for 1 hr at room temperature. AP activity was detected with CDP Star substrate (Tropix, Bedford, MA) in glow luminometer. The response data of the series of diluted reference material were fitted and the curve was used for quantification of the tumor extracts. The value of the undiluted SKBr3 extract was denominated 1 U/ml.

2.8. Immunoassay of P-Akt level. Black ninety-six–well microtiter plates (Nunc Black MaxiSorp Surface; Nalgen Nunc International, Rochester, NY) were coated with anti-Akt/PKB, PH domain antibody, clone SKB1, Cat. No. 05-591 (Upstate Biotechnology, Lake Placid, NY) at a concentration of 3 mg/mL of coating buffer in a volume of 100 µl/well and kept at 4°C overnight. For determination of phosphorylated Akt tumor extracts were prepared as described above in the presence of Na_3VO_4. Before sample applications, the coated microtiters were washed five times with 200 µl/well of washing buffer and then blocked for 2 hours at room temperature with 250 µl of blocking buffer. The blocked wells were washed five times with 300 ml of blocking buffer and then 100 ml of the diluted tumor membrane extracts or reference material was added to the wells and incubated overnight at 4°C. As a reference for each assay, a cell extract of MCF-7 cells, stimulated as described above, was used. For use in ELISAs, the MCF-7 cell membrane extract was sequentially diluted with sample dilution buffer at ratios of 1X, 0.75X, 0.5X, 0.25X, 0.125X, and 0.025X, and then 100 µl aliquots were incubated on each microtiter plate together with the tumor tissue extracts and blanks (containing only dilution buffer). After incubation of the samples and references, the wells were washed five times with 300 µl washing buffer at room temperature to eliminate unbound particles. Biotinylated detection antibody (Phospho-Akt (Ser473), clone 4E2, Cell Signaling Technologies, Beverly, MA) was added to the wells and incubated for 2 hrs at room temperature. Complex was detected with horseradish peroxidase (HRP)-conjugated streptavidin, diluted in conjugate diluent for 1 hr at room temperature. HRP activity was detected with SuperSignal WestPico substrate (Pierce) in glow luminometer. The response data of the series of diluted reference material were fitted and the curve was used for quantification of the tumor extracts. The value of the undiluted MCF-7 extract was denominated 1 U/ml.

2.9. Immunoassay of pan-Y and S phosphorylation of ShcA. Black ninety-six–well microtiter plates (Nunc Black MaxiSorp Surface; Nalgen Nunc International, Rochester, NY) were coated with anti-Shc antibody, Cat. No. 06-203 (Upstate Biotechnology, Lake Placid, NY) at a concentration of 3 mg/ml of coating buffer in a volume of 100 µl/well and kept at 4°C overnight. For determination of phosphorylated ShcA tumor extracts were prepared as described above in the presence of Na_3VO_4. Before sample applications, the coated microtiters were washed five times with 200 µl/well of washing and then blocked for 2 hours at room temperature with 250 µl of blocking buffer. The blocked wells were washed five times with 300 µl of blocking buffer and then 50 µl of the diluted tumor membrane extracts or reference material was added to the wells and incubated overnight at 4°C. As a reference for each assay, a s cell extract of SKBr3 cells, stimulated as described above, was used. For use in ELISAs, the SKBr3 cell membrane extract was sequentially diluted with sample dilution buffer at ratios of 1X, 0.75X, 0.5X, 0.25X, 0.125X, and 0.025X, and then 100 µl aliquots were incubated on each microtiter plate together with the tumor tissue extracts and blanks (containing only dilution buffer). After incubation of the samples and references, the wells were washed five times with 300 µl washing buffer at room temperature to eliminate unbound particles. Biotinylated

PhosphoSerine Antibody Q5, Cat. No. 37430 or PhosphoThreonine Antibody Q7, Cat. No. 37420 (Qiagen, Chatsworth, CA) or alkaline phosphatase (AP) conjugated PhosphoTyrosine antibody PY20, Cat. No. 03-7722 (Zymed Laboratories Inc., South San Francisco, CA) was added to the wells and incubated for 2 hrs at room temperature. Complex was detected with AP-conjugated streptavidin, diluted in conjugate diluent for 1 hr at room temperature. AP activity was detected with CDP Star substrate (Tropix, Bedford, MA) in glow luminometer. The response data of the series of diluted reference material were fitted and the curve was used for quantification of the tumor extracts. The value of the undiluted SKBr3 extract was denominated 1 U/ml.

2.10. RNA extraction, quantification and quality check (performed by OncoScore, AG staff). Total RNA was extracted from powdered tumor tissue aliquots using the RNeasy Mini Kit (Qiagen) according to the manufacturer's recommendations. Purified RNAs were stored at -70°C. RNA was quantified and checked for quality by the Bioanalyzer 2100 and the RNA 6000 Nano LabChip-Kit (Agilent Technologies) following the manufacturer's recommendations.

2.11. cDNA Synthesis (performed by OncoScore, AG staff). 1μg of good quality total RNA was reverse transcribed in a final volume of 20μl. Final concentrations were 10mM DDT, 1μg of hexamer primers, 2 U of MMLV Reverse Transcriptase (Invitrogen), 40 U of RNasin (Promega), 0.5mM of each dNTP (Promega), 1x reaction buffer. RNA, hexamer primers and dNTPs were incubated for 5' at 65°C, cooled for 5' at 4°C and subsequently treated for 10' at 25°C before the rest of the reaction mix was added. Reverse transcription was carried out at 37°C for 50' followed by 5' at 42°C. As soon the reaction was completed 180μl of H$_2$O were added and cDNAs were stored at -20°C until use.

2.12. Primer design (performed by OncoScore, AG staff). Primers were designed using the Primer Express™ v2.0 software (Applied Biosystems, Forster City, CA). All primer sequences were blasted against the dbEST and nr databases to confirm their specificity and from GeneScan Europe (Freiburg, Germany). All primer sets were chosen to work at the same conditions (AT=60°C) and to be cDNA specific. All primer sets were tested in Qrt-PCR reactions on universal human reference RNA (Stratagene, La Jolla, CA).

2.13. Quantitative real-time RT-PCR (performed by OncoScore, AG staff). PCR was performed in 40 cycles on a ABI Prism 7000 using the 2x SYBR Green I Master Mix (Applied Biosystems, Forster City, CA) in a final volume of 25µl. After PCR, all amplicons were submitted to a temperature ramping and analyzed for their melting points and sequenced. Relative quantification (µCt) was obtained by normalization with ribosomal 18S. Qrt-PCR results were expressed in arbitrary Units per µg reverse transcribed RNA (U/µg rt-RNA). All three markers were measured in parallel. Inter-assay variation was monitored on Human Universal Standard RNA (Stratagene) using GSTP (glutathione S-transferase-P1) as reference gene. The assays were highly reproducible with a coefficient of variation less than 0.15 among the different run.

2.14. EGF binding assay for EGFR quantification. Binding assays were carried out in 96-well plates in a total volume of 250 µl. All plasticware was preblocked with a 1% BSA solution before use. Final reaction concentrations were 20 mM HEPES, pH 7.5, and 0.1% BSA containing 20 µg of membrane preparation in the presence or absence of crocidolite asbestos or riebeckite. The final concentration of asbestos in the binding reactions was calculated to be approximately equivalent to 5 µg/cm^2 of cell membrane. An excess of unlabeled EGF (final concentration 4 µM) was added to the tubes in which nonspecific binding was to be measured. Stock solutions were divided into aliquots into microwells containing increasing amounts ^{125}I-EGF to initiate the binding reactions. The binding assay was conducted at room temperature for 1 h. The reaction mixtures were diluted in 1 ml of 20 mM HEPES (pH 7.5) and 0.1% BSA. The dilute reaction mixtures were filtered to separate receptor-bound ^{125}I-EGF from free ligand by use of a vacuum filtration manifold. Reaction mixtures were filtered through glass microfiber filters (Whatman, Maidstone, UK; Schleicher & Schuell, Keene, NH) that had been wetted with HEPES-BSA. After filtration, each filter was washed several times with HEPES-BSA. Filters were then counted in a gamma spectrometer. Samples were analyzed in duplicate. Total binding corresponded to conditions in the absence of unlabeled EGF, across which increasing concentrations of ^{125}I-EGF were titrated. Nonspecific binding corresponded to conditions in which there was an excess of unlabeled EGF. Specific binding was calculated by subtracting nonspecific binding from total binding.

2.14. Sample preparation for SDS-PAGE. The supernatant of a cell lysate obtained as described above was mixed 1/1 with 2x protein sample buffer (125 mM Tris Base, 20% (v/v) Glycerol, 2% (w/v) SDS, 0.02% Bromphenol blue, 2% mertcaptoethanol), heat for 5 min. at 95°C. Store at -80C.

2.15. SDS-PAGE. Protein gel buffers were cast as described by Laemmli et al. (276) using buffers described in material section. The BioRad Modular Mini-Protean II Electrophoresis Gel System was used for protein electrophoresis.

2.16. Western-Blotting. Western blot analysis. The proteins were resolved on a 10% sodium dodecyl sulfate (SDS)-polyacrylamide gel as described above, transferred to Immobilon (Millipore) membrane using buffers described below, incubated for 1 h at RT in BLOTTO, and subsequently probed with the primary antibody for 3 to 4 h at room temperature or overnight at 4°C. The primary antibody diluted 200-2000x in BLOTTO was used to detect the relevant protein. Following incubation with the primary antibody, the blots were rinsed with PBST (0.05% Tween-20 in PBS) and probe for 1 h at room temperature with a species-specific secondary antibody conjugated to horseradish peroxidase (HRP), the blot was washed again with PBST and then with PBS and subjected to ECL detection. Membranes were stripped using stripping buffer and reprobed with another antibodies, when necessary.

2.17. ECL detection. HRP- conjugated secondary antibodies we detected using SuperSignal West Pico Chemiluminescent Substrate. Two substrate components were mixed at a 1:1 ratio to prepare the substrate Working Solution. Blot was incubated for 5 minutes in Working Solution. Excess reagent was drained and blot was covered with clear plastic wrap. Blot was then exposed to X-ray film.

2.18. Computing. The SofMax program was used for curve-fitting and marker amount quantification (U/ml). The Prism and S+ programs were used for statistical analysis and drawing of graphs.

2.19. Statistics. Statistical significance of the association between P-Akt and other dichotomous variables (e.g., nodal status), was assessed by Mann-Whitney test. The continuous variable function of CLISA-determined P-Akt values was first tested for prognostic significance by univariate Cox regression analysis against patient relapse rate. A cut-off or prognostic threshold value was searched for by means of classification and regression trees analysis (CART) (273, 274). Survival probabilities were calculated by the Kaplan-Meier method and compared by means of logrank analysis (275). Spearman rank correlation (r_s) was calculated to assess associations between continuous markers (e.g. P-ErbB2 and EGFR protein expression levels).

3. Immunoassay development

3.1. Introduction.

A direct way to assess specific activation of kinases or other signaling proteins is by analysis of their phosphorylation (80, 166, 167). Analyses of protein phosphorylation have previously relied primarily on slide-based immunohistochemistry (IHC) or Western blot analyses with enzyme-conjugated phosphospecific antibodies (168, 169). Typically these techniques are time-consuming and have low sample throughput. Moreover, in the case of Western blot analysis, accurate quantitation can also be very cumbersome, while in case of IHC semiquantitative scoring systems are used, which are subject to interobserver interpretation error and which prevent statistical analysis of the values as a continuous variable function. On the other hand, assays performed on frozen or fresh tumor tissue produce more consistent results than those subject to variable types of tissue fixation and archival storage of paraffin-embedded sections. It is often the case, that, different IHC antibodies different staining patterns and intensities depending on the type of tissue fixative and the length of slide storage.

Enzyme immunoassay (EIA) measurement of phosphorylated protein levels averts the potential antigen damage associated with fixation, embedding, and uncontrolled storage and yields a highly reproducible continuous value results. However a disadvantage of such methods is the requirement for relatively large quantities of fresh or frozen tumor tissue. Furthermore, the two-site (sandwich) EIA assays ensure increased specificity as compared to one-antibody assays such as IHC and Western Blotting

We have developed a rapid, sensitive, and high-throughput kinase phosphorylation assays termed a two-site chemoluminescence-linked immunosorbent assay (CLISA) to analyze protein phosphorylation in well-characterized primary breast cancer samples. Since chemoluminometric detection provides higher sensitivity for the detection of antigen/antibody complex as compared to colorimetric assays, we have expected detection of phosphoproteins in the present study to be extremely sensitive as well as specific and precise.

3.2. Results.

Seven assays have been developed: PY1248 ErbB2, pan-PY ErbB2, pan-PS ErbB2, pan-PT ErbB2, PS473-Akt, pan-PY Shc and pan-PY Shc assays. Since all the assays have undergone exactly the same validation procedures, below will be presented an example of the assay development, the development of a pan-PY Shc assay.

3.2.1. Western Blotting. Western Blotting, as a standard assay, was used in order to check the performance of newly purchased antibodies. Proteins from SkBr3, BT474, MCF-7 and BT20 cell lysates were separated using 10% SDS PAGE gel and transferred to PVDF membrane (Milipore). Western Blot was visualized using SuperSignal WestPico ECL detection reagents (Fig. 8).

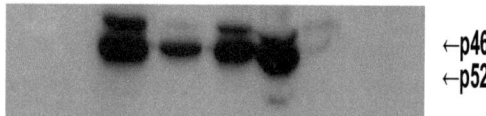

Figure 8. Western blotting with the anti-ShcA antibody was performed in human cell lines (MCF-7, BT474, SKBr3 and THP-1 cells) as described under "Experimental Procedures." ShcA protein was detected as ~50-kDa bands (see *arrows*).

3.2.2. Dose-response curve, LOD and LOQ. Different antibody pairs were tested using series of standard dilutions (undiluted, 5X, 25X, 125X, 625X, 3125X, 15625X, 78125X, 390625X and 1953125X diluted standard). As shown in Fig. 9 dilution curve could be drawn only if the assay was functional.

Limit of detection (LOD) and limit of quantification (LOQ) was calculated from the dose-response curve. LOD was calculated as the doses of signals at two-fold standard deviation above blank level, LOQ at ten-fold standard deviation above blank level. LOD of PY-ShcA assay was 0.00027 U/ml , LOQ - 0.062 U/ml

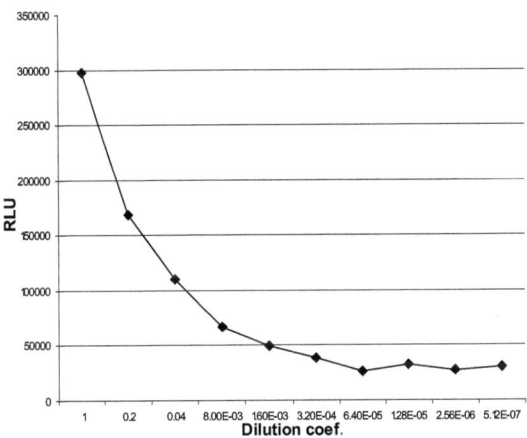

Figure 9. Dose-response curve.

3.2.3. Specificity. In order to test the specificity of the assay and the extent of nonspecific binding, different controls were used, omitting one reagent required for the assay. As seen in Fig. 10, assays lacking either analyte or detecting antibody or substrate or conjugate or coating antibody, show at least 8 fold lower RLU than the assay containing all the components. Species-specific antibody directed against the coating antibody was used as a positive control, and shows 12 fold higher RLU than the normal assay.

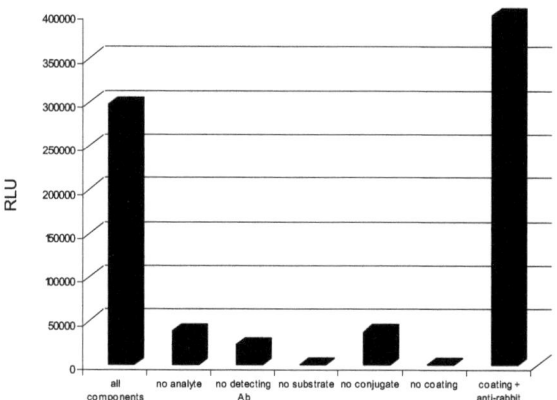

Figure 10. Controls used in the assay in order to role out nonspecific binding. Assays lacking one component, show only a background RLU as compared to the complete assay.

3.2.4. Phospho-specificity. In order to test whether the assay was specific for the pohosphorylated form of ShcA, BT20 and BT474 cell lines were serum starved for 48 hrs and either lysed untreated or treated with EGF or Na_3VO_4. Lysates were normalized for total protein content and measured using newly developed assay. In BT20 and BT47 cell lines EGF treatment increased RLU 1.3 fold and 3.3 fold respectively, and Na_3VO_4 treatment – 1.7 fold and 10.5 fold respectively (Fig. 11), suggesting that the assay indeed detects phosphorylated form of ShcA.

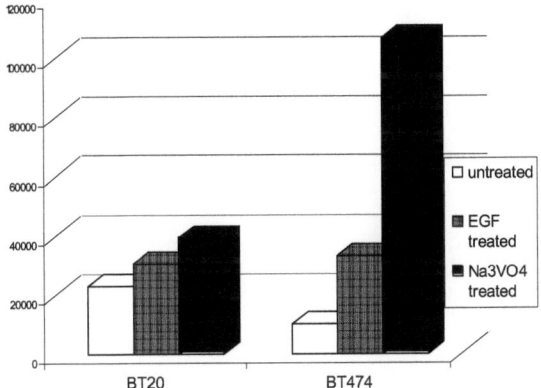

Figure 11. The proof of a phospho-specificity of the assay. EGF or Na_3VO_4 treatment increased RLU readings, confirming the phospho-specificity of the assay.

3.2.5. Standard curve. The value of the undiluted SKBr3 extract was denominated 1 U/ml.

The response data of the series of diluted reference material (1, 0.75, 0.5, 0.25, 0.125 and 0.025 U/ml) and blanks (only dilution buffer: 0 U/ml) were fitted and the standard curve (Fig. 12) was used for quantification of the tumor extracts.

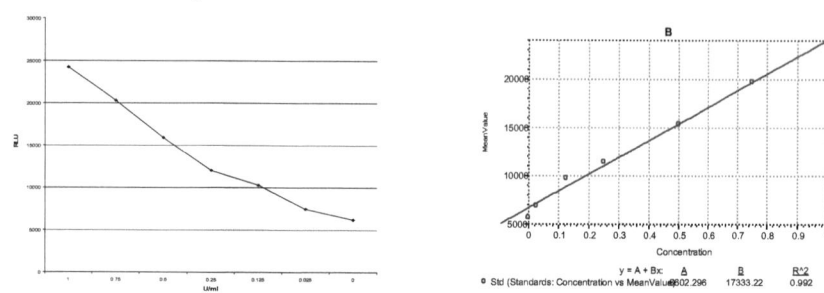

Figure 12. The standard curve: Excel graph (A) and SoftMax graph (B).

3.2.6. The raw data. After the measurement cancer sample values were calculated by fitting the standard curve, using SoftMax Pro software. The resulting raw result (U/ml) (Fig. 12) were divided by total sample protein levels (mg/ml) in order to obtain absolute PY ShcA content in given tumor sample (U/mg of total protein contents).

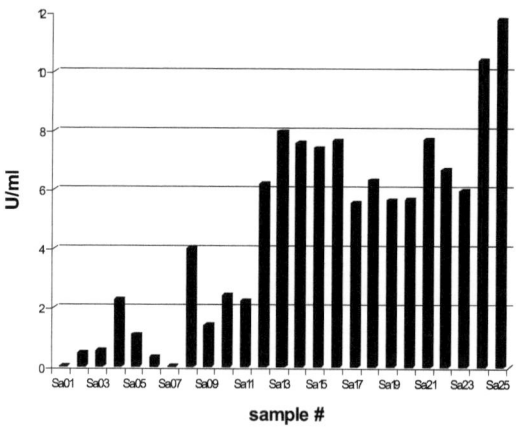

Figure 13. The mean result of PY ShcA detection in 25 samples after fitting the standard curve.

3.2.7. Final result. We applied a novel immunoassay in primary breast tumor tissue samples of 157 breast cancer patients. For the entire collective CLISA quantified PY ShcA levels ranged from 0 to 32.33 U/mg with a median of 3.997 U/mg (mean 4.59 U/mg). The distribution of PY ShcA levels in these samples is shown in Fig. 14.

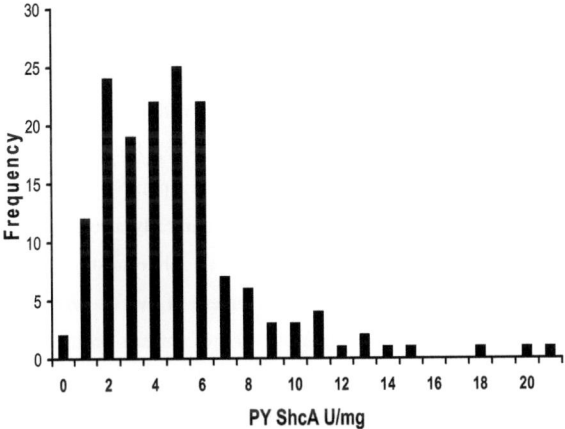

Figure 14. Histogram showing distribution of PY ShcA levels in 157 primary breast cancer samples.

4. The role of Y1248 ErbB2 phosphorylation in primary breast cancer.

4.1. Introduction.

ErbB2 is amplified and overexpressed in 10-30% of breast cancers (170-172). It's amplification and overexpression have been associated with poor prognosis or response to anticancer therapies (173). An inverse relationship between estrogen receptor and ErbB2 expression has been reported in breast cancer, with overexpression associated with decreased sensitivity to endocrine therapy (172; 174-175). Therapy based on a humanized monoclonal anti-ErbB2 antibody (trastuzumab/Herzeptin™) has been beneficial in metastatic patients (27). Trastuzumab is the first clinically available oncogene-targeted therapeutic agent for treatment of solid tumors. First-line trastuzumab in combination with chemotherapy resulted in a 25% improvement in overall survival compared with chemotherapy alone, but still a maximum of 40% of patients profit. Moreover, different groups have postulated a possible resistance to SERM in breast cancer patients whose tumors overexpress ErbB2 (174; 176). Therefore, more precise identification of the patients that are more likely to respond to such treatment is necessary in order to improve survival and quality of life.

ErbB2 (Her2/Neu) is a member of the epidermal growth factor receptor (EGFR) family of receptor tyrosine kinases, which includes EGFR(HER1, ErbB1), ErbB3(HER3) and ErbB4(HER4) (177-180). This family of receptors regulates cell proliferation through activation of downstream signal transduction cascades. The ErbB2 molecule is composed of an extracellular ligand binding domain, a transmembrane region and an intracellular region, which contains tyrosine kinase domain and a carboxy tail with five major autophosphorylation sites (80). Ligand binding initiates the formation of homo- and heterodimeric complexes with other EGFR family members into which ErbB2 is recruited as preferential dimerisation partner (84). This promotes autophosphorylation of specific tyrosine residues, which provide docking sites for a variety of proteins involved in the activation of downstream intracellular signaling cascades (41). In vitro overexpression of ErbB2 can result in receptor autophosphorylation and activation even in the absence of the activating ligands (181-184). Activating mutations have been described experimentally, but none has been detected in human tumors (185-188). It has also been suggested that shedding of the extracellular domain may represent an alternative activation mechanism of ErbB2 (189-190). Amplification or overexpression of ErbB2 has been demonstrated in many epithelial tumors, such as breast, prostate and non small cell lung cancers (191-193).

However, ErbB2 gene amplification or overexpression may not necessarily reflect the functional status of the receptor and it has been hypothesized that ErbB2 could be activated to different degrees in tumors expressing the same amounts of the receptor. So far, several studies have investigated the phosphorylation of ErbB2 in breast tumors and its possible association with poor prognosis or prediction of therapies (171; 194-198). The primary aim of the present study was to revalidate our previous results (171) on the prognostic role of PY1248 ErbB2in patients overexpressing ErbB2 as well as in patients with intermediate levels of ErbB2. A secondary aim was to determine possible associations between phosphorylation of ErbB2, and the expression levels of EGFR family members as well as estrogen receptor (ER) and progesterone receptor (PgR).

4.2. Results.

4.2.1. Tumor and Patient Characteristics. A new collective of 70 well characterized primary breast tumors cryopreserved (-80°C) in our tumor bank, were selected according to the ErbB2 expression levels previously routinely detected by ELISA at the time of surgery. Thirty tumors overexpressed ErbB2, the remaining 30 had intermediate (100-260 ng/mg) levels. During a median follow-up time of 55 months (range 30 to 89 months) 14 (20%) patients died because of the disease and 24 (34%) developed a relapse. Fifty (71%) cases were ER positive tumors (>20 fmol/mg protein), 38 (54%) PgR positive tumors (>20 fmol/mg protein). Thirty three (47%) cases were node-positive tumors and 34 (48%)

of the tumors were grade III. Most of the tumors (42) were of medium size (T2), 18 (26%) T1 and 10 (14%) T3. Median age of patients was 56 (range 28-82)(Tables 1, 2).

Table 1. Clinicopathological Characteristics of the Patients

Feature	Number of patients (%)
Patients enrolled	70
Histology type	
Ductal	48 (69)
Lobular	11 (16)
Other	11 (16)
Tumor size	
T1	18 (26)
T2	42 (60)
T3-4	10 (14)
Lymph-node status	
Node-negative	33 (47)
Node-positive	37 (53)
Histopathological grade	
I + II	27 (39)
III	34 (48)
Not analyzed	9 (13)
Estrogen receptor (ER)	
Positive (>20 fmol/mg)	50 (71)
Negative (≤ 20 fmol/mg)	20 (29)
Progesterone receptor (PgR)	
Positive (>20 fmol/mg)	38 (54)
Negative (≤ 20 fmol/mg)	32 (46)

Table 2. Features of the Entire Primary Breast Tumor Collective

Factor	Median	Mean	Range
Age	56	58	28-82
ER	72 fmol/mg	139 fmol/mg	2-625 fmol/mg
PgR	28 fmol/mg	129 fmol/mg	0-1250 fmol/mg
ErbB2	308 ng/mg	299 ng/mg	121-654 ng/mg

4.2.2. Distribution of PY1248 ErbB2 Levels in Breast Cancer Membrane Fractions. For the entire collective of 70 tumors (whose clinicopathological features are described in Table 1), CLISA quantified ErbB2 levels ranged from 0 to 127.304 U/mg with a median of 1.0955 U/mg (mean 7.49 U/mg) (Figure 15). Increased levels of tyrosine 1248 phosphorylated (PY1248) ErbB2 were found in ErbB2 - overexpressing tumors, median 2.89 U/mg (mean 12.41 U/mg) as compared to low-ErbB2-expressing tumors, median 0.39 U/mg (mean 0.93 U/mg). 27% (8 of 30) of low-ErbB2-expressing tumors contain PY1248 ErbB2 above clinically the relevant threshold value (see below), and 68% (27/40) of ErbB2 - overexpressing tumors.

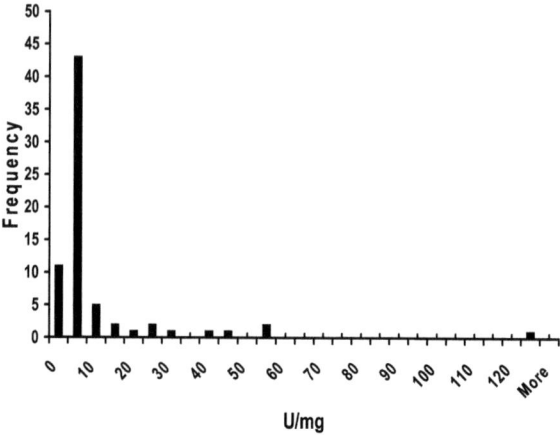

Fig 15. Histogram showing distribution of CLISA-determined PA1248 erbB2 expression levels in 70 primary breast cancer samples.

4.2.3. Prognostic Significance of CLISA-Determined Threshold Value of Tumor PY1248 ErbB2 Levels. To determine whether there were any significant biologic difference between groups of tumors containing increased levels of PY1248 ErbB2 quantified by CLISA, particularly those distributed above or below the median PY1248 ErbB2 value (1.0955 U/mg), patient outcome was evaluated with respect to different tumor ErbB2 phosphorylation levels measured in 70 samples. As shown by the Kaplan-Meier plots of Fig. 16, patients with high PY1248 ErbB2 tumors not only had significantly reduced DFS ($P < 0.005$) but also had reduced OS ($P=0.08$).

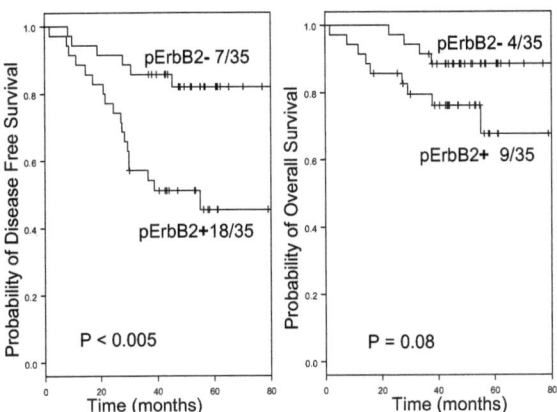

Fig 16. Kaplan-Meier survival curves stratified by CLISA-determined PY1248 ErbB2 levels. Patients whose tumors contain high levels of PY1248 ErbB2 (above median value) show a significantly worse outcome. Significant differences in both in DFS and in OS curves.

4.2.4. High PY1248 ErbB2 Levels are Associated With Reduced Tumor ER and PgR Content.
Only 14% (5/35) of high PY1248 ErbB2 containing tumors were ER- negative (> 20 fmol/mg), and only 31% (11/35) PgR- negative (> 20 fmol/mg). In contrast, 43% (15/35) of low PY1248 ErbB2 containing tumors were ER- negative, and 63% (22/35) PgR-negative (Figure 17). Analysis of quantitative levels of PY1248 ErbB2 revealed inverse correlation with quantitative mRNA as well as protein expression levels of ER and PgR (Figure 18, Table 3).

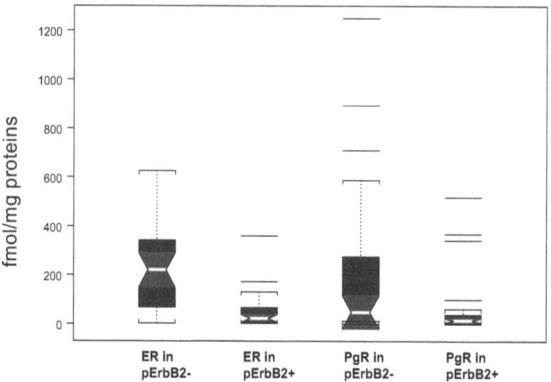

Fig 17. Box-plot showing the dramatically reduced tumor ER and PgR protein levels in association with high levels of PY1248 ErbB2 (above median value).

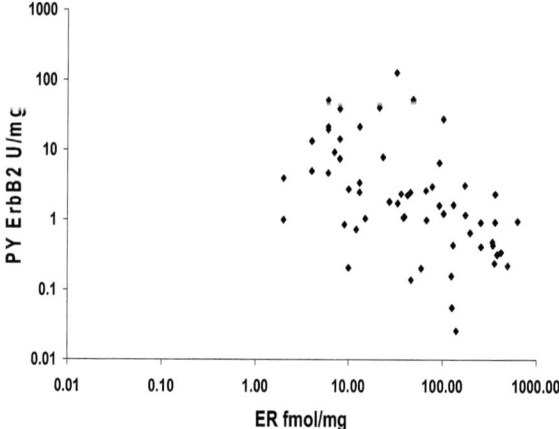

Fig 18. Scatter plot of PY1248 ErbB2 versus ER expression levels.

4.2.5. Correlation of PY1248 ErbB2 Levels and mRNA expression levels of EGFR family members. Spearman rank correlation analysis of quantitative PY1248 ErbB2 levels and quantitative mRNA expression levels of EGFR family receptors (detected by OncoScore, AG staff) was carried out. Comparison of ErbB family expression on mRNA level with ErbB2 phosphorylation revealed significant correlation with ErbB2 (r_s=0.62, p<0.001) and EGFR (r_s=0.26, p=0.05) (Table 3) and inverse correlation with ErbB3 (r_s=-0.22, p=0.08) and ErbB4 (r_s=-0.47, p=0.0005) (Table 3).

Table 3. Spearman rank correlation analysis of quantitative PY1248 ErbB2 levels and quantitative levels of EGFR family receptors, ER and PgR.

PY1248 ErbB2 Correlation With:	Spearman correlation coefficients with respect to mRNA expression levels	Spearman correlation coefficients with respect to protein expression levels
ER	-0.54 (p<0.001)	-0.67 (p<0.001)
PgR	-0.46 (p<0.001)	-0.45 (p<0.001)
EGFR	0.26 (p=0.05)	0.43 (p<0.005)
ErbB2	0.62 (p<0.001)	0.53 (p<0.0005)
ErbB3	-0.22 (p=0.08)	n.d.
ErbB4	-0.47 (p=0.0005)	n.d.

4.2.6. Correlation of PY1248 ErbB2 Levels and protein levels of EGFR family members. Spearman rank correlation analysis of quantitative PY1248 ErbB2 levels and quantitative protein levels of EGFR and ErbB2 receptors was carried out. Significant correlation was found between EGFR levels, detected by EGF binding assay (r_s=0.43, p<0.005) and PY1248 ErbB2 (Fig. 19, 20 and Table 3). Good agreement between PY1248 and ErbB2 levels, detected by EIA was found (r_s=0.53, p<0.0005) (Table 3).

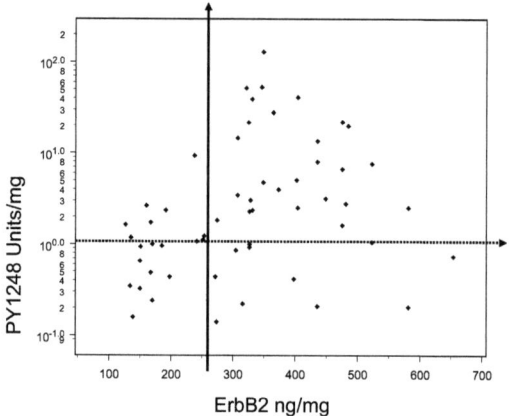

Fig 19. Scatter plot of PY1248 ErbB2 versus ErbB2 expression levels.

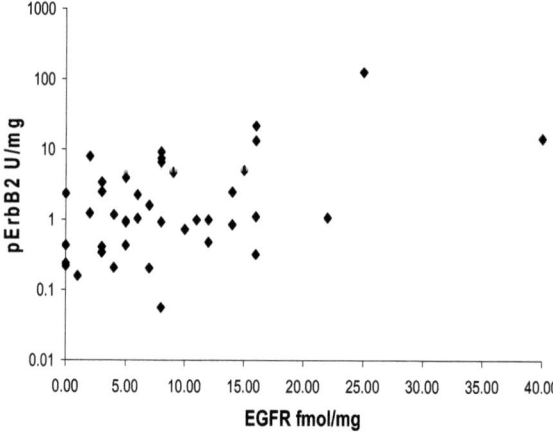

Fig 20. Scatter plot of PY1248 ErbB2 versus EGFR expression levels.

4.2.7. Correlation of mRNA Levels and Protein Levels of EGFR family members, ER and PgR. Spearman rank correlation analysis of quantitative mRNA levels and quantitative protein levels of EGFR, ErbB2, ErbB3, ErbB4, ER and PgR was carried out in order to see whether the correlations between protein expression and mRNA acceptable . Good agreement was found for all factors used in this study (Table 4).

Table 4. Spearman rank correlation analysis of quantitative mRNA levels and quantitative protein levels of EGFR, ErbB2, ER and PgR.

Factor	Spearman correlation coefficient between mRNA and protein expression levels
ER	0.84 (p<0.001)
PgR	0.72 (p<0.001)
EGFR	0.48 (p<0.005)
ErbB2	0.59 (p<0.001)

4.3. Discussion.

This is the first study analyzing the activation of ErbB2 in breast tumors in relation to the expression of EGFR family of receptor tyrosine kinases as well as other clinical – pathological variables and survival. We aimed to confirm our previous data on the prognostic role of PY1248 ErbB2 in patients overexpressing ErbB2 and on the other hand to assess the prognostic value of PY1248 ErbB2 in patients with intermediate levels of ErbB2. We also aimed to reveal probable relations between phosphorylation of ErbB2, and the expression levels of EGFR family members as well as hormone receptors.

In this study, PY1248 ErbB2was considered positive in 50% of the tumors and it was not associated with nodal status or tumor size, but we did find a strong association with EGFR and ErbB2 expression and strong inverse correlation with ER, PgR and ErbB4 expression. ErbB4 mRNA expression has been shown to be associated with good prognosis in a number of studies (199-201), supporting the inverse correlation with P-ErbB2. There was also weak inverse correlation between PY1248 ErbB2 and ErbB3 expression. Since it has been observed that ErbB3 collaborates with ErbB2 in driving breast tumor cell proliferation (202), the negative association between PY1248 ErbB2 and ErbB3 was unexpected. Nevertheless, ErbB3 mRNA expression has been shown to be associated with good prognosis in several studies (199-200), as well as in our collective (data not shown). Relationship between protein levels of ErbB3 and ErbB2 as well as phosphorylation of both factors could possibly better explain these discrepancies.

It has been suggested by several studies that ErbB2 in a heterodimer with EGFR, is involved in signaling pathways that are required for a human breast cancer cell to become metastatic (203-204). Our data further supports the importance of EGFR/ErbB2 heterodimer in breast cancer.

We conclude that the phosphorylation of ErbB2 could be a factor to consider together with other tumor characteristics in predicting relapses of breast cancer.

5. The role of Akt phosphorylation in primary breast cancer

5.1. Introduction.

Akt/protein kinase B (PKB) is a serine/threonine protein kinase involved in mediating various biological responses including inhibiting apoptosis and stimulating cell proliferation (205-206). So far three mammalian isoforms of this enzyme are known: Akt1/PKBα, Akt2/PKBβ, and Akt3/PKBγ (205). Akt1 is the predominant form in most tissues, whereas Akt2 is highly expressed in insulin-responsive tissues and Akt3 is relatively high expressed in brain and testis (206). Akt1 was first discovered as cellular homolog of a viral oncogene v-Akt causing leukemia in mice (207). Akt is known to be activated by phosphoinositol 3 phosphates (PIP3), the products of PI3K activity (208), downstream of a wide range of receptors. Among these receptors the c-Met receptor (209), epidermal growth factor receptor (EGFR) family (210), fibroblast growth factor receptor (211), insulin growth factor receptor (IGF) (212), and platelet-derived growth factor receptor (PDGFR) (213) are found. In addition, Akt activity can be regulated by the PTEN tumor suppressor gene, which negatively regulates PIP3 levels (214). Akt1 is activated by phosphorylation, after PIP3 binding, on two critical residues: threonine 308 (T308) and serine 473 (S473). Similar residues are highly conserved in Akt2 and Akt3 (205-206). Several studies have found Akt2 to be amplified or overexpressed on mRNA level in primary tumors and cell lines (215-217). Overexpression of Akt2 protein was also described in a number of human carcinomas, such as colon, pancreatic and breast tumors (218-220). However, Akt2 activation by phosphorylation may have more important prognostic value than Akt2 amplification or overexpression. On the other hand, phosphorylated Akt1 and Akt3 could be as important as phosphorylated Akt2.

To this date, several groups have investigated the phosphorylation of Akt in breast, prostate, colon and pancreatic tumors using immunohistochemical (IHC) methods (218; 221-226). However, phosphorylation structures may be disturbed by formalin imbedding rendering specific antigen sites difficult to be recognized. Moreover, IHC gives only semiquantitative results limiting statistical analysis. In contrast, enzyme immunoassays (EIA) have the advantage to yield highly reproducible results of continuous values.

In the following investigation we applied a novel CLISA which allows to detect phosphorylated Akt1, 2 and 3 in well characterized primary breast tumor tissue samples of 156 breast cancer patients.

5.2. Results.

5.2.1. Tumor and Patient Characteristics. A collective of 156 cryopreserved (-80°C) and well characterized primary breast tumors was selected from our tumor bank according to the ErbB2 protein expression levels routinely detected at time of surgery in order to better investigate interactions between P-Akt and ErbB2. All patients underwent primary surgery before January 1996. 67 patients experienced a disease recurrence within the median follow-up time of 57 months (range 27-88 months). Sixty six patients (42%) were nodal negative and 90 (58%) nodal positive. Forty tumors (26%) were estrogen receptor (ER) negative. None of the patients received neoadjuvant therapy. Details on patients and tumor characteristics are summarized in Table 5.

Table 5. Clinicopathological characteristics of the patients.

Feature	Number of patients (%)
Patients enrolled	156
Age	
<40	12 (8)
40-60	85 (54)
>60	59 (38)
Histology type	
Ductal	109 (70)
Lobular	17 (11)
Other	30 (19)
Tumor size	
T1	49 (31)
T2	90 (58)
T3-4	17 (11)
Lymph-node status	
Node-negative	66 (42)
Node-positive	90 (58)
Histopathological grade	
I + II	57 (37)
III	86 (55)
Not analyzed	13 (8)
Estrogen receptor (ER)	
Positive (>20 fmol/mg)	116 (74)
Negative (≤ 20 fmol/mg)	40 (26)
Progesterone receptor (PgR)	
Positive (>20 fmol/mg)	85 (54)
Negative (≤ 20 fmol/mg)	71 (46)

5.2.2. Distribution of P-Akt Levels in Breast Cancer Cytosolic Fractions. CLISA quantified P-Akt levels ranged from 0 to 1.084 U/mg with a median of 0.1665 U/mg (mean 0.199 U/mg). As shown in figure 21 P-Akt levels have a normal distribution. There was no correlation between the levels of P-Akt and ErbB2. Moreover, no correlation was found between the levels of P-Akt and nodal status, estrogen receptor (ER) status, tumor size nor differentiation staging.

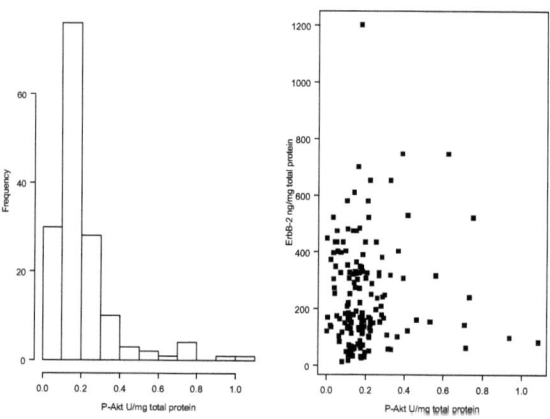

Figure 21. (A) Histogram showing distribution of CLISA-determined P-Akt expression levels in 156 primary breast cancer samples. P-Akt levels ranged from 0 to 1.084 U/mg with a median of 0.1665 U/mg. (B) Scatter plot of P-Akt versus ErbB2 expression levels. No correlation was found between the levels of P-Akt and ErbB2.

5.2.3. Prognostic Significance of P-Akt Levels. The prognostic value of P-Akt was investigated with respect to relapse free survival in the overall collective. Univariate Cox-regression revealed a weak correlation between P-Akt expression levels and relapse-free survival ($p<0.05$). In order to visualize this result, we calculated an optimal cut-off value for P-Akt by CART and plotted Kaplan-Meier survival curves stratifying the patients according to the P-Akt levels of their tumors (Figure 22). Out of the 21 patients (14% of all collective) with tumors expressing P-Akt levels higher than 0.31 U/mg, 14 patients relapsed within the observation time with a disease-free survival rate of 33% (CI: 17.5-61%) significantly ($p < 0.01$) worse as compared to those tumors with lower P-Akt levels (DFS rate of 60%, CI: 67-40%).

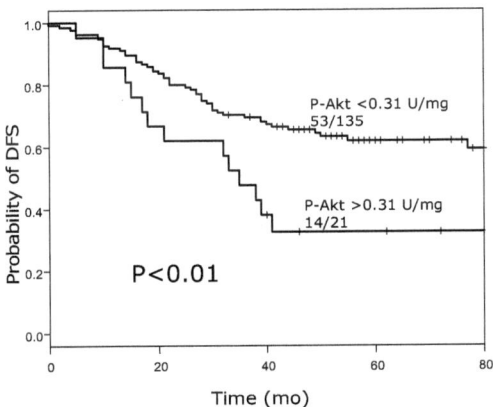

Figure 22. Total collective. Kaplan-Meier survival curves stratified by P-Akt levels. Patients whose tumors contain high levels of P-Akt show a significantly worse outcome in respect to DFS (p<0.01).

5.2.4. Prognostic Significance of P-Akt in a subset of ErbB2-overexpressing tumors.

Although no correlation was found between P-Akt and ErbB2 expression, the prognostic impact of P-Akt in the subset of ErbB2-overexpressing tumors was more evident. The Kaplan-Meier curves in Fig. 23a reveal that patients outcome decreases significantly when tumors express P-Akt levels higher than the median value (p=0.005) and even more when P-Akt expression levels exceed the third quartile value (p<0.001), indicating a cumulative prognostic impact of these 2 biomarkers (Fig. 23b).

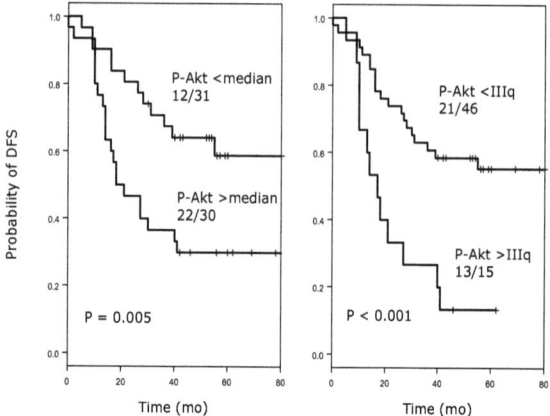

Figure 23. Subset of patients with ErbB2 overexpressing tumors. Kaplan-Meier survival curves stratified by median (A) and last quartile (B) values of P-Akt. Patients whose tumors express high levels of P-Akt show a significantly worse outcome in respect to DFS.

5.2.5. Correlation of P-Akt Levels and mRNA expression of proliferation markers.

The quantitative P-Akt protein levels were correlated by Spearman rank with the quantitative mRNA expression levels of more than 60 biomarkers involved in different biological processes (detected by OncoScore, AG staff), such as proliferation, hormone dependency, apoptosis, growth factors and angiogenesis. P-Akt expression levels were found to correlate only with proliferation makers. A good correlation was found with respect to thymidylate synthase (TS) ($r_s=0.38$, $p<0.001$) (Figure 24), whereas the positive correlations found between P-Akt and thymidine kinase 1 (TK1) ($r_s=0.23$, $p<0.01$), survivin ($r_s=0.22$, $p<0.01$), topoisomerase II alpha (TOPO II alpha) ($r_s=0.19$, $p<0.05$), and transcription factor E2F ($r_s=0.22$, $p<0.01$) were less prominent.

The P-Akt levels were also correlated by Spearman rank with the quantitative mRNA expression levels of more than 40 000 genes using DNA microarrays (carried out by Dr. Patrick Urban, using home-made DNA microarrays, ISREC, Laussane, Switzerland).

Correlation between numerous genes involved in Cell Cycle / Prolifereation and P-Akt was reconfirmed, as seen in Table 6.

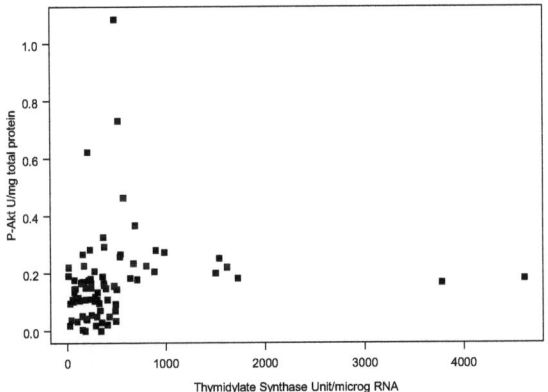

Figure 24. Scatter plot of P-Akt versus TS mRNA expression levels. The strong significant correlation (rs=0.38, p<0.001) was found.

5.2.6. Genes Associated with P-Akt. We have also carried out the comparison of gene expression in P-Akt positive vs P-Akt negative samples by Wilcoxon test using DNA microarrays. The significant difference in expression levels of more than 25 genes was found, as shown in Table 7. Some other relevant genes, such as Akt1 and ErbB2 were close to being significant.

Table 6. Spearman rank correlation analysis of quantitative P-Akt levels with 25 Top Genes from Array.

Spearman correlation (rs)	Gene	ChromLocation	Description
0.51	TYMS	18p11.32	thymidylate synthetase
0.48	HMGB2	4q31	high-mobility group box 2
0.38	KIF22	16p11.2	kinesin family member 22
0.36	CENPF	1q32-q41	centromere protein F, (mitosin)
0.34	ZNF354A	5q35.3	zinc finger protein 354A
0.34	RANBP17	5q34	RAN binding protein 17
0.33	RAD54L	1p32	RAD54-like (S. cerevisiae)
0.32	CDC2	10q21.1	cell division cycle 2, G1 to S and G2 to M
0.32	UHRF1	19p13.3	ubiquitin-like, containing PHD and RING finger domains, 1
0.32	ZNF200	16p13.3	zinc finger protein 200
0.32	EZH2	7q35-q36	enhancer of zeste homolog 2 (Drosophila)
0.32	CBX1	17q	chromobox homolog 1 (HP1 beta homolog Drosophila)
0.31	DDX11	12p11	DEAD/H (Asp-Glu-Ala-Asp/His) box polypeptide 11
0.31	CGI-01	1q24-q25.3	CGI-01 protein
0.31	FEN1	11q12	flap structure-specific endonuclease 1
0.31	TP53BP2	1q42.1	tumor protein p53 binding protein, 2
0.31	SFRS1	17q21.3 q22	splicing factor, arginine/serine rich 1
0.31	KIF20A	5q31	kinesin family member 20A
0.31	LUC7L	16p13.3	LUC7-like (S. cerevisiae)
0.30	PCNA	20pter-p12	proliferating cell nuclear antigen
0.30	STMN1	1p36.1-p35	stathmin 1/oncoprotein 18
0.30	C16orf34	16p13.3	chromosome 16 open reading frame 34
0.30	CDC2	10q21.1	cell division cycle 2, G1 to S and G2 to M
0.30	MGC24665	16p13.2	hypothetical protein MGC24665
0.29	RAD51	15q15.1	RAD51 homolog (RecA homolog, E. coli) (S. cerevisiae)

Table 7. Genes Associated with pAkt positive vs. pAkt negative. Thirty Top Genes from Array.

p value	Gene Name	Chrom. Location	Description
0.0012	CHGB	20pter-p12	chromogranin B (secretogranin 1)
0.0078	LLGL2	17q24-q25	lethal giant larvae homolog 2 (Drosophila)
0.0118	C1orf29	1p31.1	chromosome 1 open reading frame 29
0.0218	APOE	19q13.2	apolipoprotein E
0.0264	THRAP1	17q22-q23	thyroid hormone receptor associated protein 1
0.0271	CYP4B1	1p34-p12	cytochrome P450, family 4, subfamily B, polypeptide 1
0.0278	ARNT2	15q24	aryl-hydrocarbon receptor nuclear translocator 2
0.0298	DHRS2	14q11.2	dehydrogenase/reductase (SDR family) member 2
0.0298	STIP1	11q13	stress-induced-phosphoprotein 1 (Hsp70/Hsp90-organizing protein)
0.0348	GRIA2	4q32-q33	glutamate receptor, ionotropic, AMPA 2
0.0358	AREG	4q13-q21	amphiregulin (schwannoma-derived growth factor)
0.0373	G22P1	22q13.2-q13.31	thyroid autoantigen 70kDa (Ku antigen)
0.042	KRT20	17q21.2	keratin 20
0.042	ZZEF1	17p13.3	zinc finger, ZZ-type with EF hand domain 1
0.0424	AGTR1	3q21-q25	angiotensin II receptor, type 1
0.0447	SLC39A6	18q12.2	solute carrier family 39 (zinc transporter), member 6
0.0458	CCL19	9p13	[e-value: 1.7e-87] chemokine (C-C motif) ligand 19
0.0481	IFI27	14q32	interferon, alpha-inducible protein 27
0.0491		NA	MRNA; cDNA DKFZp686F01160 (from clone DKFZp686F01160)
0.05	C7	5p13	complement component 7
0.05	SDC4	20q12	syndecan 4 (amphiglycan, ryudocan)
0.052	KRT16	17q12-q21	keratin 16 (focal non-epidermolytic palmoplantar keratoderma)
0.0538	MGC5395	11q12-q13	hypothetical protein MGC5395
0.0546	SDC4	20q12	syndecan 4 (amphiglycan, ryudocan)
0.0549	PSAP	10q21-q22	prosaposin (variant Gaucher disease and variant metachromatic leukodystrophy)
0.0592	ERBB2	17q11.2-q12	v-erb-b2 erythroblastic leukemia viral oncogene homolog
0.0605	SERPINB5	18q21.3	serine (or cysteine) proteinase inhibitor, clade B (ovalbumin), member 5
0.0627	TPD52L1	6q22-q23	tumor protein D52-like 1
0.0652	TCF20	22q13.3	transcription factor 20 (AR1)
0.0703	AKT1	14q32.32	v-akt murine thymoma viral oncogene homolog 1

5.3. Discussion

This investigation demonstrates clearly that breast cancer patients with high levels of Akt phosphorylation (13.5%) have a significantly worse relapse free prognosis as shown in Fig. 21. No correlation was found between levels of P-Akt, ErbB2 expression levels, nodal Status, ER/PgR status, tumor size and/or tumor differentiation.

Similar results have been previously reported using IHC as standard method. Perez-Tenorio *et al.* (221) demonstrated that bad prognosis was correlated with increased phosphorylation of Akt in endocrine therapy-treated patients, whereas Schmitz *et al.* (223) observed prognostic relevance of P-Akt in node-negative breast cancer and Stal *et al.* (222) showed that Akt activation is associated with decreased local benefit from radiotherapy.

Our study is the first report on P-Akt assessment by EIA using a phospho-specific antibody in breast cancer cytosols of cryopreserved samples, since it was previously pointed out that assays performed on frozen tumor tissue allow more precise and quantitative results (227). In contrast to the semi-quantitative and partly subjective IHC data, tumor marker expression levels assessed by quantitative EIA measurements are more reliable with respect to sensitivity and reproducibility. Moreover, EIA results obtained from fresh frozen tissue extracts avoid the potential antigen damage due to formalin fixation, paraffin embedding and uncontrolled storage. Furthermore, the two-site (sandwich) CLISA assay used ensures increased specificity as compared to one-antibody assays such as IHC and Western Blotting. In addition chemoluminometric detection provides high sensitivity for the detection of antigen/antibody complex.

We assumed to find a correlation between ErbB2 and P-Akt, since ErbB2 has been implicated in the Akt activation (228). Nevertheless, Akt can be also activated by various other receptor tyrosine kinases (209-213), by G protein-coupled receptors (229), and by other molecules often deregulated in breast cancer. Loss of PTEN activity in breast cancer (230) is accompanied by increased expression of the activated form of Akt. Based on these findings it is clear that Akt can be activated by ErbB2 as well as by other stimuli. Therefore Akt activation has not to be necessarily correlated with ErbB2 overexpression though we have shown that the prognostic significance of P-Akt is increased if congruent with ErbB2 overexpression.

Of interest is the correlation of P-Akt with mRNA expression levels of tumor proliferation markers we observed in this study. Akt is known to promote cell cycle progression by modulating the expression (231) and stabilization of cyclin D1 (232), which in turn activates the E2F transcription factor. Our results demonstrate a significant correlation of P-Akt with E2F transcription factor expression of as well as the expression of E2F-regulated genes, namely TS, TK, survivin and TOPO IIalpha (Fig. 25).

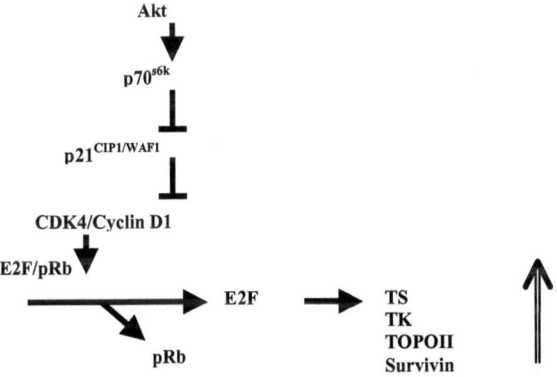

Fig. 25. Schematic representation of Akt-induced cell cycle progression. Akt signaling pathway modulates the stabilization of cyclin D1, which in turn activates the E2F transcription factor as well as the expression of E2F-regulated genes.

In conclusion, we were able to reconfirm that high levels of P-Akt is a marker of poor prognosis (decreased DFS), although Akt Phosphorylation is not associated with the nodal status of the patients or with other tumor characteristics including the ErbB2 status.

Moreover, the prognostic value of P-Akt levels increased in the subset of patients with ErbB2-overexpressing tumors. In addition, the phosphorylation of Akt was found to be associated with the expression of tumor proliferation factors. Based on these results, we propose to investigate the predictive value of P-Akt with respect to Herceptin and topoisomerase inhibitor therapy. Lastly, we suggest combination therapies with Akt inhibitors in a subset of well-characterized human breast tumors.

6. Analysis of ErB2, ShcA and Akt phosphorylation in primary breast cancer

6.1. Introduction.

Molecular analysis of human tissue specimens is performed for diagnostic and prognostic purposes. Morphologic examination provides complex visual information about cell and tissue characteristics, whereas molecular analysis provides valuable data about the expression of individual proteins or genes. The most common method for extracting molecular information from human tissues is to visualize protein levels in tissue either by immunohistochemistry or EIA approaches. This analysis, however, remains limited in its ability to elucidate the function of such proteins. Protein function is regulated not only by transcription and translation, but also by such posttranslational modifications as phosphorylation, glycosylation, and prenylation. The most reversible of these modifications is phosphorylation, a reaction that involves transfer of a nucleotide phosphate to recipient tyrosine or serine/threonine amino acid. This mechanism lies behind most pathways cell signal transduction and is thus fundamental both normal physiology and disease. Given these considerations, the number of proteins in the activated (phosphorylated) state may be more important biologically than the total number of proteins present. We therefore assumed that protein phosphorylation (indicative of signaling activity), either in addition to or rather than protein overexpression data, might improve the prognostic value.

ErbB2 is a 185 kDa oncoprotein (p185), which is overexpressed in 25–30% of invasive breast cancers. Intracellular phosphorylated tyrosine residues of the ErbB2 molecule function as high-affinity binding sites for SH2 domain containing proteins, which link the receptor to intracellular signal transduction pathways such as the Ras-Raf-MAPK and the PI3K-Akt pathways. Both are believed to be key elements in the regulation of cell proliferation and survival (233). In this context, phosphorylation of the Y1248, which is supposed to constitute main autophosphorylation site of ErbB2, is a key event for downstream signalling (234), followed by the phosphorylation of other autophosphorylation sites, namely Y1023, Y1139, Y1196 and Y1121/1122. Receptor tyrosine kinase signalling may be physiologically terminated either by homologous receptor desensitization to the continued presence of ligand, or else via heterologous receptor transmodulation by other ligand-activated tyrosine kinases. The best desensitization studies has been performed on the EGFR. On each side of the EGFR catalytic domain lies a desensitization domain: the juxtamembrane region containing the Thr654 PKC site (235-237), and the carboxyterminal SH2-like domain containing Ser1046/7 the calcium/calmodulin-dependent kinase II (CamKII) site (238). Like the EGFR, desensitization of the ErbB2 tyrosine kinase is mediated through phosphorylation of serine and threonine residues within the intracellular part of the receptor (101).

Akt/PKB is a central signaling serine/threonine kinase that has been implicated in the genesis or progression of numerous human tumors, because it regulates many of the key effector molecules involved in apoptosis, anoikis and cell cycle progression (205). Activation of the Akt pathway suppresses apoptotic response, undermines cell cycle control and selectively enhances the production of key growth and survival factors. Upon stimulation, Akt is recruited to the plasma membrane through the binding of its N-terminal pleckstrin homology (PH) domain to PIP_3, a lipid product of PI3K (208). Akt is then activated by phosphorylation on two residues: Thr-308 in the activation loop and Ser-473 in the hydrophobic motif of the C-terminal tail.

The adapter protein Shc is a SH2 containing proto-oncogene involved in growth factor signaling. In addition to the ubiquitously expressed ShcA, two other isoforms, ShcB and ShcC exist, which are expressed in neuronal cells ShcA in vivo. Based on molecular and mouse knock-out studies major role for ShA leading to MAPK activation has been established (120-121). ShcA is expressed as three splice forms of about 46, 52 and 66 kDa. Within the CH domain of ShcA, three tyrosine-phosphorylation sites have been identified. The Y239/Y240 twin tyrosines have been linked to c-Myc activation (239-240); however, the mechanism by which Shc leads to c-Myc activation is not known. Y317 tyrosine has been established in leading to MAP kinase activation through Grb2 and Sos (241). The p66Shc possesses an additional CH2 domain that contains a serine phosphorylation site that has been implicated in oxidative stress signaling (242).

We applied novel two-site CLISA assays, which allowed us to detect Y, S and T phosphorylated ErbB2, Y and S phosphorylated ShcA and S473 phosphorylated Akt1, 2 and 3 in well characterized primary breast tumor tissue samples of 153 breast cancer patients.

6.2. Results.

6.2.1. Tumor and Patient Characteristics. A collective of 153 cryopreserved (-80°C) and well characterized primary breast tumors was selected from our tumor bank according to the ErbB2 and ER protein expression levels routinely detected at time of surgery. 80 (52%) tumors overexpressed ErbB2, the remaining 73 (48%) had intermediate (100-260 ng/mg) levels. Ninety eight tumors (64%) were estrogen receptor (ER) negative, the remaining 55 (36%) had low (20-50 fmol/mg) ER levels. All patients underwent primary surgery before January 1996. Seventy seven (50%) patients experienced a disease recurrence within the median follow-up time of 47 months (range 24-87 months). Seventy five patients (49%) were nodal negative and 58 (38%) nodal positive. None of the patients received neoadjuvant therapy. Details on patients and tumor characteristics are summarized in Table 8. Due to the fact that tumors we selected for this study were mostly aggressive tumors (high ErbB2, low ER, nodal positive, relapsing tumors), it was impossible to analyze the prognostic value of the molecules studied.

Table 8. Clinicopathological characteristics of the patients.

Feature	Number of patients (%)
Patients enrolled	153
Age <40	10 (6.5)
40-60	88 (57.5)
>60	55 (36)
Histology type	
Ductal	104 (68)
Lobular	9 (6)
Other	40 (26)
Tumor size	
T1	39 (25)
T2	87 (57)
T3-4	24 (16)
Unknown	3 (2)
Lymph-node status	
Node-negative	58 (38)
Node-positive	75 (49)
Unknown	20 (13)
Histopathological grade	
I + II	49 (32)
III	57 (37)
Not analyzed	47 (31)
Estrogen receptor (ER)	
Positive (>20 fmol/mg)	55 (36)
Negative ($\leq$ 20 fmol/mg)	98 (64)
Progesterone receptor (PgR)	
Positive (>20 fmol/mg)	40 (26)
Negative ($\leq$ 20 fmol/mg)	113 (73)

6.2.2. Distribution of PY, PS and PT ErbB2 Levels in 153 Breast Cancer Membrane Fractions. CLISA quantified PY ErbB2 levels ranged from 0 to 1000 U/mg with a median of 3.72 U/mg (mean 54.49 U/mg). Increased levels PY ErbB2 were found in ErbB2 - overexpressing tumors, median 6.39 U/mg (mean 86.28 U/mg) as compared to low-ErbB2-expressing tumors, median 2.06 U/mg (mean 19.66 U/mg), which reconfirmed the PY1248 ErbB2 result (Fig. 26). CLISA quantified PS ErbB2 levels ranged from 0 to 198.64 U/mg with a median of 0 U/mg (mean 5.85 U/mg). No PS ErbB2 was detected in ErbB2 - overexpressing tumors. In a subpopulation of low-ErbB2-expressing tumors, median of PS ErbB2 levels was 5.7 U/mg (mean 12.18 U/mg) (Fig 27). CLISA quantified PT ErbB2 levels ranged from 0 to 1000 U/mg with a median of 0 U/mg (mean 0.15 U/mg). As in the case of PS ErbB2, no PT ErbB2 was detected in ErbB2 - overexpressing tumors. In a subpopulation of low-ErbB2-expressing tumors, median of PT ErbB2 levels was 0 U/mg (mean 0.32 U/mg). (Fig. 28).

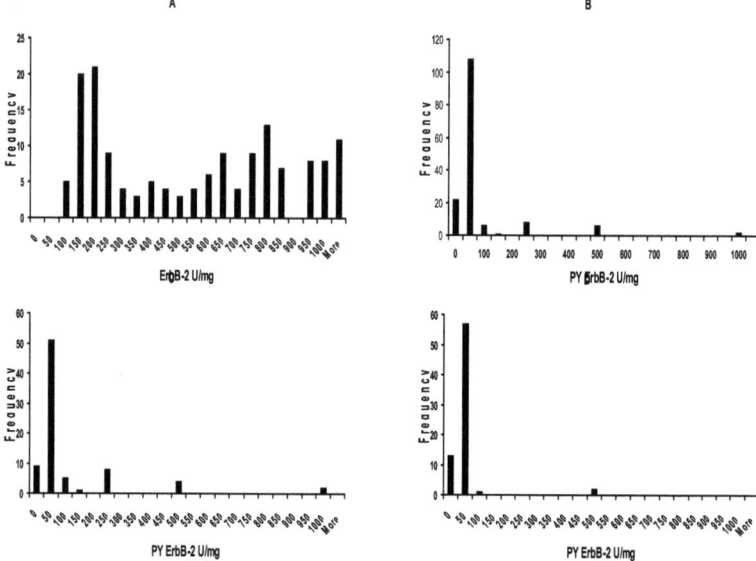

Fig 26. Histogram showing distribution of EIA-determined ErbB2 expression levels in total collective of 153 primary breast cancer samples (A), CLISA-determined PY1248 ErbB2 expression levels in total collective (B), and subpopulation of erbB2-overexpressing tumors (C) and subpopulation of low-erbB2 expressing tumors (D).

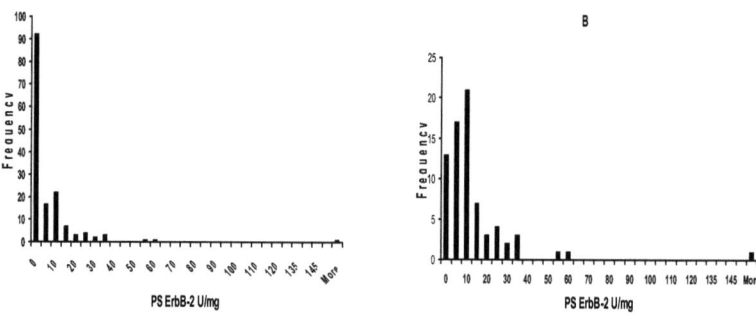

Fig 27. Histogram showing distribution of CLISA-determined PS 1248 ErbB2 expression levels in total collective of 153 primary breast cancer samples (A) and subpopulation of low-erbB2 expressing tumors (B).

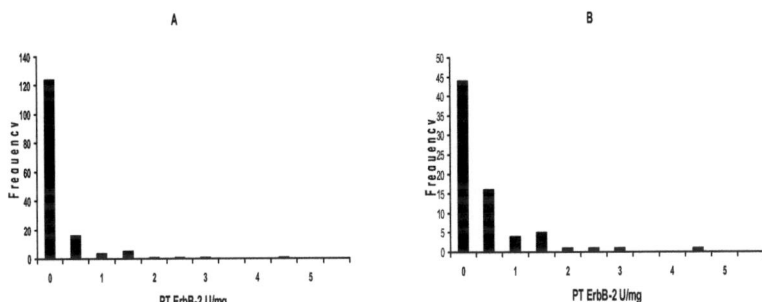

Fig 28. Histogram showing distribution of CLISA-determined PT 1248 ErbB2 expression levels in total collective of 153 primary breast cancer samples (A) and subpopulation of low-erbB2 expressing tumors (D).

6.2.3. Correlation of PY, PS and PT ErbB2 Levels and ErbB2 expression levels.

Spearman rank correlation analysis of quantitative PY, PS and PT ErbB2 levels and quantitative ErbB2 expression levels was carried out. Comparison of ErbB2 expression on protein level with ErbB2 phosphorylation revealed significant correlation with PY ErbB2 (r_s=0.33, p<0.001) and EGFR (r_s=0.26, p=0.05) and inverse correlation with PS ErbB2 (r_s=-0.53, p<0.01) (Figure 29, Table 9). There was no significant correlation between ErbB2 expression and PT ErbB2 levels.

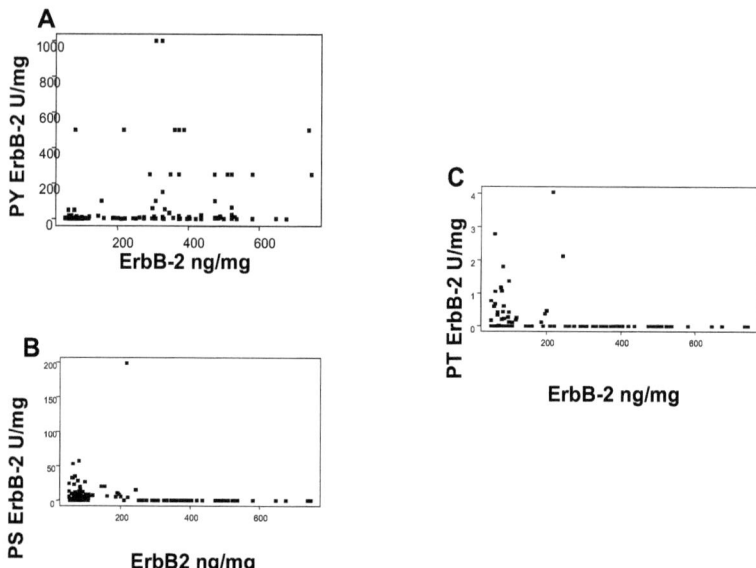

Fig 29. Scatter plot of PY ErbB2 versus ErbB2 expression levels (A), PS ErbB2 versus ErbB2 expression levels (B) and PT ErbB2 versus ErbB2 expression levels (C). . The strong significant correlation (rs=0.33, p<0.01) was found between PY ErbB2 and ErbB2 expression and significant inverse correlation (rs=-0.53, p<0.01) was found between PS ErbB2 and ErbB2 expression.

6.2.4. Distribution of PY and PS ShcA Levels and PS473 Akt in Breast Cancer Membrane Fractions. CLISA quantified PY SchA levels ranged from 0 to 32.33U/mg with a median of 4.05 U/mg (mean 4.63 U/mg) and PS SchA levels ranged from 0 to 19.49 U/mg with a median of 3.76 U/mg (mean 4.27 U/mg) (Figure 30).

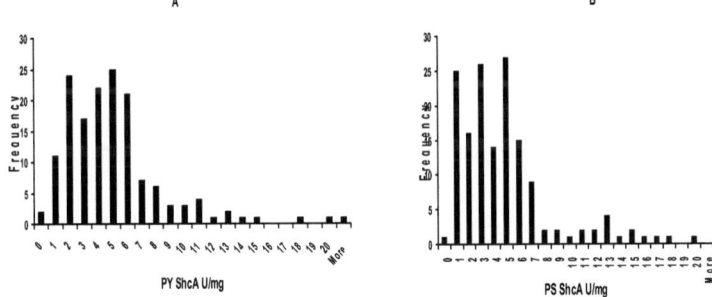

Fig 30. Histogram showing distribution of CLISA-determined PY SchA (A) and PS SchA (B) levels in total collective of 153 primary breast cancer samples.

6.2.5. Correlations between PY, PS and PT ErbB2, PY and PS ShcA and PS473 Akt. The quantitative PY, PS and PT ErbB2, PY and PS ShcA and PS473 Akt protein levels were correlated by Spearman rank. Good correlation was found between PS ErbB2 and PT ErbB2(r_s=0.74, $p<0.01$), PS ShcA and PY SchA (r_s=0.88, $p<0.01$), PT ErbB2 and ER(r_s=0.34, $p<0.01$), and PT ErbB2 and PgR(r_s=0.39, $p<0.01$), (Figure 31, Table 9), whereas the positive and/or inverse correlations found between other factors were less prominent or even insignificant (Table 9).

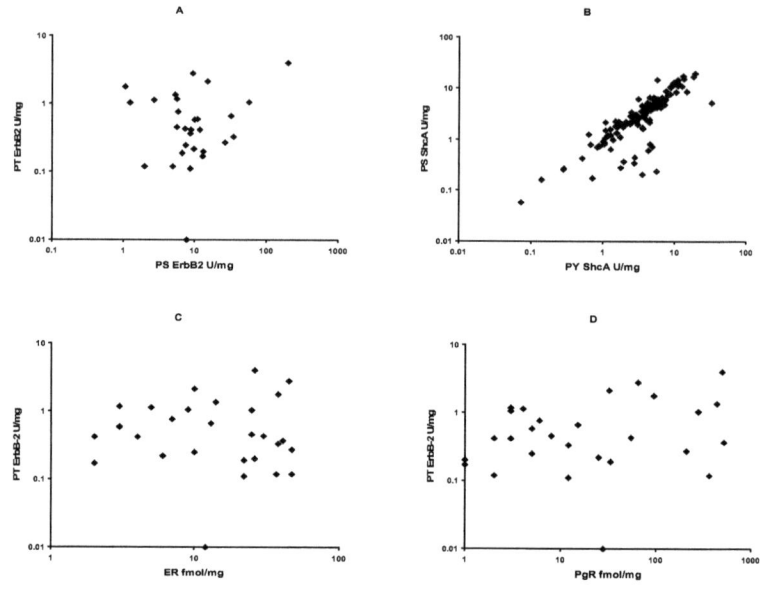

Fig 31. Scatter plot of PS ErbB2 versus PT ErbB2 levels (A), PS ShcA versus PT ShcA levels (B), PT ErbB2 versus ER expression levels (C) and PT ErbB2 versus PgR expression levels (D).

Table 9. Spearman rank correlation analysis of quantitative PY, PS and PT ErbB2, PY and PS ShcA and PS473 Akt.

	PY ErbB2	PS ErbB2	PT ErbB2	P Akt	PY ShcA	PS ShcA
ErbB2	$r_s=0.33$	$r_s=-0.53$	n.s.	n.s.	n.s.	$r_s=0.18$
ER	n.s.	$r_s=0.20$	$r_s=0.34$	n.s.	n.s.	n.s.
PgR	n.s.	$r_s=0.27$	$r_s=0.39$	n.s.	n.s.	n.s.
PY ErbB2		n.s.	$r_s=0.26$	$r_s=0.20$	n.s.	n.s.
PS ErbB2	n.s.		$r_s=0.74$	n.s.	n.s.	n.s.
PT ErbB2	$r_s=0.26$	$r_s=0.74$		$r_s=0.18$	$r_s=0.27$	$r_s=0.21$
P Akt	$r_s=0.20$	n.s.	$r_s=0.18$		$r_s=-0.27$	$r_s=-0.28$
PY ShcA	n.s.	n.s.	$r_s=0.27$	$r_s=-0.27$		$r_s=0.88$
PS ShcA	n.s.	n.s.	$r_s=0.21$	$r_s=-0.28$	$r_s=0.88$	

6.3. Discussion

Our study is the first report on assessment of pan S, T and Y phosphorylation of ErbB2 and S and T phosphorylation by EIA using a phospho-specific antibody in breast cancer cytosols of cryopreserved samples.

It has been reported that differentially phosphorylated ErbB2 isoforms vary in such functional features as ligand-dependent catalytic activity (243-244), heterodimerization (245-246), down-regulation kinetics (246-247), and intracellular trafficking (245, 248). Having shown in this report that patterns of site-specific phosphorylation differ between cancer patients, it is reasonable to infer that ErbB2 may not have an identical role in all tumors. However, quite clear prevalence of PY ErbB2 in ErbB2 overexpressing tumors could be seen both from the results this study as well as the PY 1248 assessment.

Consistent with this is preliminary data showing an association between ErbB2 kinase activity, and PY ErbB2 immunoreactivity (243, 249).

Interestingly, PS and PT ErbB2 was not detected in ErbB2 overexpressing tumors, but quite high levels could be detected in low ErbB2 expressing tumors. This data is consistent with the fact that desensitization of the ErbB2 kinase activity is mediated through phosphorylation of serine and threonine residues of the receptor via PKC (248). Moreover PT and PY ErbB2 strongly correlates with ER and PgR protein levels, further inferring the "opposing" function of serine and threonine phosphorylation. PT and PS ErbB2 levels correlated very well with each other, suggesting similar function and/or regulation.

It has been reported, that all EGFR family members have docking sites for ShcA binding either via SH2 or PTB domains (250), therefore it was not very surprising that ShcA phosphorylation did not correlate significantly neither with ErbB2 expression nor with it's phosphorylation. However, significant, although weak correlation was found with PT ErbB2 levels.

Unfortunately, clinical outcome in this analysis could not be carried out due to the selection of the collective used in present study. The criteria used to select the collective (high or moderate ErbB2 levels and low ER levels), resulted in selection of highly aggressive subset of tumors, which made survival analysis impossible to be performed. In order to further continue analysis in order to investigate the prognostic role of pan-phosphorylations of ErbB2 and ShcA different collective of tumors should be used, selected upon different criteria or just a collective of randomly selected samples.

7. Discussion and Perspectives

Since the value of ErbB2 (HER-2/*neu*) analysis for prognosis or prediction has been controversial, our goal was to determine whether ErbB2 signaling improved the prognostic power of ErbB2 analyses. Prognostic significance of Y1248-phosphorylated ErbB2 has been previously investigated by several groups including ours (), in both fixed and unfixed tissues using a variety of methods. We have also shown that PY1248 ErbB2 is detectable in only in a minority of ErbB2–overexpressing breast cancers.

In the present study, we found site-specific receptor phosphorylation to be more prevalent among ErbB2–positive cases (18%) than among ErbB2–negative cases (5%). In the entire study population PY1248 ErbB2 data provided more valuable prognostic information than ErbB2 expression data.

Our study population was not part of any particular clinical protocol and they were not uniformly treated. This excludes analyses of interactions between PY1248 ErbB2 and therapy.

Because of the expected relationship of phosphorylation and PY1248 ErbB2 positivity with high-level overexpression in general, it could be considered that PY1248 ErbB2 positivity could simply be a surrogate for high-level overexpression, especially due to the fact that in previous studies in patients which expressed low and moderate levels of ErbB2, PY1248 ErbB2 immunopositivity was not observed. However, using particularly sensitive CLISA assay, we could detect PY1248 ErbB2 in patients which expressed moderate levels of PY1248 ErbB2. This observation was confirmed in a different study, using pan-PY antibody. Unfortunately predictive value of pan-PY ErbB2 as well as pan-PS and pan-PT ErbB2, could not be carried out in the collective used for this study and therefore has to be repeated in a different, randomly selected collective. On the other hand, the number of patients used for a PY1248 study was small (n=70), reducing the power of the study and making firm conclusions difficult to draw. Therefore PY1248 ErbB2 should be assessed in a larger subset of well characterized breast tumors (n>100).

ErbB2 testing is most commonly applied because of its value as a predictive factor, either for consideration for trastuzumab treatment or because of interactions between

ErbB2 and other chemotherapeutic agents. Thus PY ErbB2 has to be evaluated in order to determine whether it is predictive of response to trastuzumab, doxorubicin, or tamoxifen.

We have also performed the first EIA study analyzing the activation of Akt in breast tumors in relation to other clinico – pathological variables and survival. We aimed to investigate whether this molecule, found in vitro to play a important role in oncogenesis showed the same significance in clinical material.

Our study demonstrates that breast cancer patients with high levels of Akt phosphorylation have a significantly worse disease free survival. We could not find any significant association of P-Akt with nodal status, tumor size, ER, PgR or other clinico – pathological features, including ErbB2 expression. On the other hand we have shown that the prognostic significance of P-Akt is increased if congruent with ErbB2 overexpression. Based on that, P-Akt should be investigated in a larger subset of ErbB2 overexpressing breast tumors (n>100), assessing not only the prognostic value but also the predictive value of P-Akt with respect to Herceptin therapy. On the other hand it remains to be elucidated phosphorylation of which Akt isoform plays the principal role. Therefore the phosphorylation and/or expression of the individual Akt isoforms should be assessed, if possible, in the same collective of tumors as total P-Akt.

Of particular interest are the correlation of P-Akt with tumor proliferation markers as well as the correlation of PY 1248 ErbB2 with EGFR and inverse correlation with ErbB4 mRNA expression levels that we observed in this study. The biology behind these correlations could be further investigated using tissue culture and/or animal models. The weak inverse correlation of ErbB2 with ErbB3 mRNA expression levels is arguing against the tissue culture – based findings and should be further investigated. Protein expression levels of ErbB3 and phosphorylation of ErbB3 should be assessed in the samples previously tested for ErbB2 expression as well as PY 1248 ErbB2 contents. Protein levels of ErbB3 (and P-ErbB3) should be evaluated both with respect to survival as well as ErbB2 (and P-ErbB2) status.

As in the case with pan-P-ErbB2 study, the predictive value of pan-PY SchA and pan-PS ShcA could not be evaluated and therefore has to be repeated in a different collective.

Although many questions remain unanswered, assuming that the above mentioned tasks can eventually be achieved, our findings raise the exciting possibility of molecular diagnostics with improved application to rational therapeutic decision making and eventually to target-specific drug development.

List of abbreviations

AP	alkaline phosphatase
ATP	adenosine triphosphate
CamK	calmoduline kinase
CART	classification and regression trees
CLISA	chemoluminescence linked immunosorbent assay
CMF	cyclophosphamide, methotrexate and 5-fluorouracil
DCIS	ductal carcinoma *in situ*
ECL	enhanced chemoluminescense
EGFR	epidermal growth factor receptor
EIA	enzyme immunoassays
ELISA	enzyme-linked immunosorbent assay
ER	estrogen receptor
Erk	extracellular signal regulated kinase
FGF	fibroblast growth factor
FGFR	fibroblast growth factor receptor,
FGFR	fibroblast growth factor receptor
FISH	fluorescent in situ hybridization
GDP	guanosine diphosphate
GSK3	glycogen synthase kinase 3
GTP	guanosine triphosphate
HM	
HRP	horseradish peroxidase
IFG-1	insulin-like growth factor 1
IGFR	insulin-like growth factor receptor
IHC	immunohistochemistry
JAK	Janus kiase
JM	juxtamembrane
kDa	kilodalton

LCIS	lobular carcinoma *in situ*,
MAPK	mitogen-activated protein kinase
MAPKAP	MAP kinase activated protein kinase
MAPKK	MAP kinase kinase
mRNA	messenger ribonucleic acid
mTOR	mammalian target of rapamycin
NFκB	nuclear factor κB
NGFR	neuronal growth factor receptor
PAI-1	plasimnogen activator inhibitor
P-Akt	phosphorylated Akt
PCR	polymerase chain reaction
PDGFR	platelet - derived growth factor receptor
PDK-1	phosphoinositol-dependent kinase 1
PgR	progesterone receptor
PH	plecstrin homology
PI3K	phosphoinositol-3 kinase
PIP3	phosphoinositol 3 phosphate
PKB	protein kinase B
PKB	protein kinase B
PLC	phospholipase C
PS	phosphorylated serine
PT	phosphorylated threonine,
PTB	phosphotyrosine binding domain,
PTEN	phosphatase and tensin homolog on chromosome 10
PTK	protein tyrosine kinase,
PY	phosphorylated tyrosine,
Q-RT-PCR	quantitative reverse transcription PCR
RLU	relative luminescence
RTK	receptor tyrosine kinase
S	serine
PAGE	polyacrylamide gel electrophoresis

SH2	Src homology 2
SDS	sodium duodecyl sulphate
STAT	signal transducer and activator of transcription
STB	Stiftung Tumorbank Basel
T	threonine
TK1	thymidine kinase 1
TOPO II alpha	topoisomerase II alpha
TS	thymidylate synthase
uPA	urokinase-type plasminogen activator
VEGFR	vascular endothelial growth factor receptor
Y	tyrosine

References

1. Baselga, J. and Norton, L. Focus on breast cancer. Cancer Cell, 1: 319-322, 2002.
2. Key, T. J., Verkasalo, P. K., and Banks, E. Epidemiology of breast cancer. Lancet Oncol, 2: 133-140, 2001.
3. Hindle, W. Breast cancer: introduction. Clin Obstet Gynecol, 45: 738-745, 2002.
4. Russo, J. and Russo, I. H. Biological and molecular bases of mammary carcinogenesis. Lab Invest, 57: 112-137, 1987.
5. Russo, J. and Russo, I. H. Toward a physiological approach to breast cancer prevention. Cancer Epidemiol Biomarkers Prev, 3: 353-364, 1994.
6. Ricketts, D., Turnbull, L., Ryall, G., Bakhshi, R., Rawson, N. S., Gazet, J. C., Nolan, C., and Coombes, R. C. Estrogen and progesterone receptors in the normal female breast. Cancer Res, 51: 1817-1822, 1991.
7. Clarke, R. B., Howell, A., Potten, C. S., and Anderson, E. Dissociation between steroid receptor expression and cell proliferation in the human breast. Cancer Res, 57: 4987-4991, 1997.
8. Russo, J., Ao, X., Grill, C., and Russo, I. H. Pattern of distribution of cells positive for estrogen receptor alpha and progesterone receptor in relation to proliferating cells in the mammary gland. Breast Cancer Res Treat, 53: 217-227, 1999.
9. Safe, S. H. Interactions between hormones and chemicals in breast cancer. Annu Rev Pharmacol Toxicol, 38: 121-158, 1998.
10. Clemons, M. and Goss, P. Estrogen and the risk of breast cancer. N Engl J Med, 344: 276-285, 2001.
11. Khan, S. A., Rogers, M. A., Khurana, K. K., Meguid, M. M., and Numann, P. J. Estrogen receptor expression in benign breast epithelium and breast cancer risk. J Natl Cancer Inst, 90: 37-42, 1998.
12. Tamoxifen for early breast cancer: an overview of the randomised trials. Early Breast Cancer Trialists' Collaborative Group. Lancet, 351: 1451-1467, 1998.

13. Esserman, L. J. New approaches to the imaging, diagnosis, and biopsy of breast lesions. Cancer J, 8 Suppl 1: S1-14, 2002.

14. Klimberg, V. S. Advances in the diagnosis and excision of breast cancer. Am Surg, 69: 11-14, 2003.

15. Nerurkar, A. and Osin, P. The diagnosis and management of pre-invasive breast disease: the role of new diagnostic techniques. Breast Cancer Res, 5: 305-308, 2003.

16. Kinne, D. W. Staging and follow-up of breast cancer patients. Cancer, 67: 1196-1198, 1991.

17. Singletary, S. E., Allred, C., Ashley, P., Bassett, L. W., Berry, D., Bland, K. I., Borgen, P. I., Clark, G. M., Edge, S. B., Hayes, D. F., Hughes, L. L., Hutter, R. V., Morrow, M., Page, D. L., Recht, A., Theriault, R. L., Thor, A., Weaver, D. L., Wieand, H. S., and Greene, F. L. Staging system for breast cancer: revisions for the 6th edition of the AJCC Cancer Staging Manual. Surg Clin North Am, 83: 803-819, 2003.

18. Fisher, E. R., Redmond, C., and Fisher, B. Histologic grading of breast cancer. Pathol Annu, 15: 239-251, 1980.

19. Noguchi, M. and Miyazaki, I. Breast conserving surgery and radiation in the treatment of operable breast cancer. Int Surg, 79: 142-147, 1994.

20. Crown, J. Evolution in the treatment of advanced breast cancer. Semin Oncol, 25: 12-17, 1998.

21. Goldhirsch, A., Colleoni, M., Coates, A. S., Castiglione-Gertsch, M., and Gelber, R. D. Adding adjuvant CMF chemotherapy to either radiotherapy or tamoxifen: are all CMFs alike? The International Breast Cancer Study Group (IBCSG). Ann Oncol, 9: 489-493, 1998.

22. Haskell, S. G. Selective estrogen receptor modulators. South Med J, 96: 469-476, 2003.

23. Miller, W. R. Aromatase inhibitors: mechanism of action and role in the treatment of breast cancer. Semin Oncol, 30: 3-11, 2003.

24. Pasqualini, J. R. and Ebert, C. Biological effects of progestins in breast cancer. Gynecol Endocrinol, 13 Suppl 4: 11-19, 1999.

25. Jonat, W. Overview of luteinizing hormone-releasing hormone agonists in early breast cancer-benefits of reversible ovarian ablation. Breast Cancer Res Treat, 75 Suppl 1: S23-26: discussion S33-25, 2002.

26. Sainsbury, R. Ovarian ablation as a treatment for breast cancer. Surg Oncol, 12: 241-250, 2003.

27. Leyland-Jones, B. Trastuzumab: hopes and realities. Lancet Oncol, 3: 137-144, 2002.

28. Ford, D., Easton, D. F., Stratton, M., Narod, S., Goldgar, D., Devilee, P., Bishop, D. T., Weber, B., Lenoir, G., Chang-Claude, J., Sobol, H., Teare, M. D., Struewing, J., Arason, A., Scherneck, S., Peto, J., Rebbeck, T. R., Tonin, P., Neuhausen, S., Barkardottir, R., Eyfjord, J., Lynch, H., Ponder, B. A., Gayther, S. A., Zelada-Hedman, M., and et al. Genetic heterogeneity and penetrance analysis of the BRCA1 and BRCA2 genes in breast cancer families. The Breast Cancer Linkage Consortium. Am J Hum Genet, 62: 676-689, 1998.

29. Elston, C. W., Ellis, I. O., and Pinder, S. E. Pathological prognostic factors in breast cancer. Crit Rev Oncol Hematol, 31: 209-223, 1999.

30. Duffy, M. J., Maguire, T. M., McDermott, E. W., and O'Higgins, N. Urokinase plasminogen activator: a prognostic marker in multiple types of cancer. J Surg Oncol, 71: 130-135, 1999.

31. Janicke, F., Prechtl, A., Thomssen, C., Harbeck, N., Meisner, C., Untch, M., Sweep, C. G., Selbmann, H. K., Graeff, H., and Schmitt, M. Randomized adjuvant chemotherapy trial in high-risk, lymph node-negative breast cancer patients identified by urokinase-type plasminogen activator and plasminogen activator inhibitor type 1. J Natl Cancer Inst, 93: 913-920, 2001.

32. Look, M. P., van Putten, W. L., Duffy, M. J., Harbeck, N., Christensen, I. J., Thomssen, C., Kates, R., Spyratos, F., Ferno, M., Eppenberger-Castori, S., Sweep, C. G., Ulm, K., Peyrat, J. P., Martin, P. M., Magdelenat, H., Brunner, N., Duggan, C., Lisboa, B. W., Bendahl, P. O., Quillien, V., Daver, A., Ricolleau, G., Meijer-van Gelder, M. E., Manders, P., Fiets, W. E., Blankenstein, M. A., Broet, P., Romain, S., Daxenbichler, G., Windbichler, G., Cufer, T., Borstnar, S., Kueng, W., Beex, L. V., Klijn, J. G., O'Higgins, N., Eppenberger, U., Janicke, F., Schmitt, M., and Foekens, J.

A. Pooled analysis of prognostic impact of urokinase-type plasminogen activator and its inhibitor PAI-1 in 8377 breast cancer patients. J Natl Cancer Inst, 94: 116-128, 2002.

33. Osborne, C. K., Yochmowitz, M. G., Knight, W. A., 3rd, and McGuire, W. L. The value of estrogen and progesterone receptors in the treatment of breast cancer. Cancer, 46: 2884-2888, 1980.

34. Pegram, M. D., Pauletti, G., and Slamon, D. J. HER-2/neu as a predictive marker of response to breast cancer therapy. Breast Cancer Res Treat, 52: 65-77, 1998.

35. Hanks, S. K., Quinn, A. M., and Hunter, T. The protein kinase family: conserved features and deduced phylogeny of the catalytic domains. Science, 241: 42-52, 1988.

36. Hanks, S. K. and Hunter, T. Protein kinases 6. The eukaryotic protein kinase superfamily: kinase (catalytic) domain structure and classification. Faseb J, 9: 576-596, 1995.

37. Taylor, S. S., Radzio-Andzelm, E., and Hunter, T. How do protein kinases discriminate between serine/threonine and tyrosine? Structural insights from the insulin receptor protein-tyrosine kinase. Faseb J, 9: 1255-1266, 1995.

38. Hubbard, S. R., Wei, L., Ellis, L., and Hendrickson, W. A. Crystal structure of the tyrosine kinase domain of the human insulin receptor. Nature, 372: 746-754, 1994.

39. Wiesmann, C., Fuh, G., Christinger, H. W., Eigenbrot, C., Wells, J. A., and de Vos, A. M. Crystal structure at 1.7 A resolution of VEGF in complex with domain 2 of the Flt-1 receptor. Cell, 91: 695-704, 1997.

40. Yayon, A., Klagsbrun, M., Esko, J. D., Leder, P., and Ornitz, D. M. Cell surface, heparin-like molecules are required for binding of basic fibroblast growth factor to its high affinity receptor. Cell, 64: 841-848, 1991.

41. Schlessinger, J. Cell signaling by receptor tyrosine kinases. Cell, 103: 211-225, 2000.

42. Hubbard, S. R., Mohammadi, M., and Schlessinger, J. Autoregulatory mechanisms in protein-tyrosine kinases. J Biol Chem, 273: 11987-11990, 1998.

43. Pawson, T. and Gish, G. D. SH2 and SH3 domains: from structure to function. Cell, 71: 359-362, 1992.

44. Songyang, Z., Shoelson, S. E., Chaudhuri, M., Gish, G., Pawson, T., Haser, W. G., King, F., Roberts, T., Ratnofsky, S., Lechleider, R. J., and et al. SH2 domains recognize specific phosphopeptide sequences. Cell, 72: 767-778, 1993.

45. Margolis, B. The PTB Domain: The Name Doesn't Say It All. Trends Endocrinol Metab, 10: 262-267, 1999.

46. Pawson, T. Protein modules and signalling networks. Nature, 373: 573-580, 1995.

47. Kundra, V., Anand-Apte, B., Feig, L. A., and Zetter, B. R. The chemotactic response to PDGF-BB: evidence of a role for Ras. J Cell Biol, 130: 725-731, 1995.

48. Schlessinger, J. and Bar-Sagi, D. Activation of Ras and other signaling pathways by receptor tyrosine kinases. Cold Spring Harb Symp Quant Biol, 59: 173-179, 1994.

49. Valius, M. and Kazlauskas, A. Phospholipase C-gamma 1 and phosphatidylinositol 3 kinase are the downstream mediators of the PDGF receptor's mitogenic signal. Cell, 73: 321-334, 1993.

50. Nomura, M., He, Z., Koyama, I., Ma, W. Y., Miyamoto, K., and Dong, Z. Involvement of the Akt/mTOR pathway on EGF-induced cell transformation. Mol Carcinog, 38: 25-32, 2003.

51. Belsches, A. P., Haskell, M. D., and Parsons, S. J. Role of c-Src tyrosine kinase in EGF-induced mitogenesis. Front Biosci, 2: d501-518, 1997.

52. Zong, C. S., Chan, J., Levy, D. E., Horvath, C., Sadowski, H. B., and Wang, L. H. Mechanism of STAT3 activation by insulin-like growth factor I receptor. J Biol Chem, 275: 15099-15105, 2000.

53. Liu, J. and Kern, J. A. Neuregulin-1 activates the JAK-STAT pathway and regulates lung epithelial cell proliferation. Am J Respir Cell Mol Biol, 27: 306-313, 2002.

54. Karin, M. and Hunter, T. Transcriptional control by protein phosphorylation: signal transmission from the cell surface to the nucleus. Curr Biol, 5: 747-757, 1995.

55. Marshall, C. J. MAP kinase kinase kinase, MAP kinase kinase and MAP kinase. Curr Opin Genet Dev, 4: 82-89, 1994.

56. Seger, R. and Krebs, E. G. The MAPK signaling cascade. Faseb J, 9: 726-735, 1995.

57. Rameh, L. E. and Cantley, L. C. The role of phosphoinositide 3-kinase lipid products in cell function. J Biol Chem, 274: 8347-8350, 1999.

58. Hunter, T. Signaling--2000 and beyond. Cell, 100: 113-127, 2000.

59. Abram, C. L. and Courtneidge, S. A. Src family tyrosine kinases and growth factor signaling. Exp Cell Res, 254: 1-13, 2000.

60. Liu, X., Brodeur, S. R., Gish, G., Songyang, Z., Cantley, L. C., Laudano, A. P., and Pawson, T. Regulation of c-Src tyrosine kinase activity by the Src SH2 domain. Oncogene, 8: 1119-1126, 1993.

61. Kypta, R. M., Goldberg, Y., Ulug, E. T., and Courtneidge, S. A. Association between the PDGF receptor and members of the src family of tyrosine kinases. Cell, 62: 481-492, 1990.

62. Prasad, K. V., Kapeller, R., Janssen, O., Duke-Cohan, J. S., Repke, H., Cantley, L. C., and Rudd, C. E. Regulation of CD4-p56lck-associated phosphatidylinositol 3-kinase (PI 3-kinase) and phosphatidylinositol 4-kinase (PI 4-kinase). Philos Trans R Soc Lond B Biol Sci, 342: 35-42, 1993.

63. Pleiman, C. M., Hertz, W. M., and Cambier, J. C. Activation of phosphatidylinositol-3' kinase by Src-family kinase SH3 binding to the p85 subunit. Science, 263: 1609-1612, 1994.

64. Arnold, S. F., Obourn, J. D., Jaffe, H., and Notides, A. C. Phosphorylation of the human estrogen receptor on tyrosine 537 in vivo and by src family tyrosine kinases in vitro. Mol Endocrinol, 9: 24-33, 1995.

65. Hamasaki, K., Mimura, T., Morino, N., Furuya, H., Nakamoto, T., Aizawa, S., Morimoto, C., Yazaki, Y., Hirai, H., and Nojima, Y. Src kinase plays an essential role in integrin-mediated tyrosine phosphorylation of Crk-associated substrate p130Cas. Biochem Biophys Res Commun, 222: 338-343, 1996.

66. Borowski, P., Kornetzky, L., Heiland, M., Roloff, S., Weber, W., and R., L. Characterization of the C-terminal domain of ras-GTPase-activating protein (ras-GAP) as substrate for epidermal growth factor receptor and p60c-src kinase. Biochem Mol Biol Int, 39: 635-646, 1996.

67. Mariner, D. J., Anastasiadis, P., Keilhack, H., Bohmer, F. D., Wang, J., and Reynolds, A. B. Identification of Src phosphorylation sites in the catenin p120ctn. J Biol Chem, 276: 28006-28013, 2001.

68. Chang, B. Y., Harte, R. A., and Cartwright, C. A. RACK1: a novel substrate for the Src protein-tyrosine kinase. Oncogene, 21: 7619-7629, 2002.

69. Ihle, J. N. Cytokine receptor signaling. Nature, 377: 591-594, 1995.

70. Valgeirsdottir, S., Paukku, K., Silvennoinen, O., Heldin, C. H., and Claesson-Welsh, L. Activation of Stat5 by platelet-derived growth factor (PDGF) is dependent on phosphorylation sites in PDGF beta-receptor juxtamembrane and kinase insert domains. Oncogene, 16: 505-515, 1998.

71. David, M., Wong, L., Flavell, R., Thompson, S. A., Wells, A., Larner, A. C., and Johnson, G. R. STAT activation by epidermal growth factor (EGF) and amphiregulin. Requirement for the EGF receptor kinase but not for tyrosine phosphorylation sites or JAK1. J Biol Chem, 271: 9185-9188, 1996.

72. Jones, F. E., Welte, T., Fu, X. Y., and Stern, D. F. ErbB4 signaling in the mammary gland is required for lobuloalveolar development and Stat5 activation during lactation. J Cell Biol, 147: 77-88, 1999.

73. Runge, D. M., Runge, D., Foth, H., Strom, S. C., and Michalopoulos, G. K. STAT1alpha/1beta, STAT 3 and STAT 5: expression and association with c-MET and EGF-receptor in long-term cultures of human hepatocytes. Biochem Biophys Res Commun, 265: 376-381, 1999.

74. Bartoli, M., Platt, D., Lemtalsi, T., Gu, X., Brooks, S. E., Marrero, M. B., and Caldwell, R. B. VEGF differentially activates STAT3 in microvascular endothelial cells. FASEB J, 17: 1562-1564, 2003.

75. Paukku, K., Valgeirsdottir, S., Saharinen, P., Bergman, M., Heldin, C. H., and Silvennoinen, O. Platelet-derived growth factor (PDGF)-induced activation of signal transducer and activator of transcription (Stat) 5 is mediated by PDGF beta-receptor and is not dependent on c-src, fyn, jak1 or jak2 kinases. Biochem J, 345: 759-766, 2000.

76. Cirri, P., Chiarugi, P., Marra, F., Raugei, G., Camici, G., Manao, G., and Ramponi, G. c-Src activates both STAT1 and STAT3 in PDGF-stimulated NIH3T3 cells. Biochem Biophys Res Commun, 239: 493-497, 1997.

77. Guren, T. K., Odegard, J., Abrahamsen, H., Thoresen, G. H., Susa, M., Andersson, Y., Ostby, E., and Christoffersen, T. EGF receptor-mediated, c-Src-dependent, activation of Stat5b is downregulated in mitogenically responsive hepatocytes. J Cell Physiol, 196: 113-123, 2003.

78. Ren, Z. and Schaefer, T. S. ErbB-2 activates Stat3 alpha in a Src- and JAK2-dependent manner. J Biol Chem, 277: 38486-38493, 2002.

79. Carpenter, G. and Cohen, S. Epidermal growth factor. J Biol Chem, 265: 7709-7712, 1990.

80. Ullrich, A. and Schlessinger, J. Signal transduction by receptors with tyrosine kinase activity. Cell, 61: 203–212, 1990.

81. Carpenter, G. and Wahl, M. I. The epidermal growth factor family. In, 1990.

82. Riese, D. J., van Raaij, T. M., Plowman, G. D., Andrews, G. C., and Stern, D. F. The cellular response to neuregulins is governed by complex interactions of the erbB receptor family. Mol Cell Biol, 15: 5770-5776, 1995.

83. Soler, C. and Carpenter, G. Guidebook to cytokines and their receptors: Oxford University Press, 1995.

84. Tzahar, E., Waterman, H., Chen, X., Levkowitz, G., Karunagaran, D., Lavi, S., Ratzkin, B. J., and Yarden, Y. A hierarchical network of interreceptor interactions determines signal transduction by Neu differentiation factor/neuregulin and epidermal growth factor. Mol Cell Biol, 16: 5276-5287, 1996.

85. Elenius, K., Paul, S., Allison, G., Sun, J., and Klagsbrun, M. Activation of HER4 by heparin-binding EGF-like growth factor stimulates chemotaxis but not proliferation. EMBO J, 16: 1268-1278, 1997.

86. Komurasaki, T., Toyoda, H., Uchida, D., and Morimoto, S. Epiregulin binds to epidermal growth factor receptor and ErbB4 and induces tyrosine phosphorylation of epidermal growth factor receptor, ErbB-2, ErbB3 and ErbB4. Oncogene, 15: 2841-2848, 1997.

87. Riese, D. J., Kim, E. D., Elenius, K., Buckley, S., Klagsbrun, M., Plowman, G. D., and Stern, D. F. Betacellulin activates the epidermal growth factor receptor and ErbB4, and induces cellular response patterns distinct from those stimulated by epidermal growth factor or neuregulin-beta. Oncogene, 12: 345-353, 1996.

88. Baulida, J., Kraus, M. H., Alimandi, M., Di Fiore, P. P., and Carpenter, G. All ErbB receptors other than the epidermal growth factor receptor are endocytosis impaired. J Biol Chem, 271: 5251-5257, 1996.

89. Downward, J., Waterfield, M. D., and Parker, P. J. Autophosphorylation and protein kinase C phosphorylation of the epidermal growth factor receptor. Effect on tyrosine kinase activity and ligand binding affinity. J Biol Chem, 260: 14538-14546, 1985.

90. Wiley, H. S., Herbst, J. J., Walsh, B. J., Lauffenburger, D. A., Rosenfeld, M. G., and Gill, G. N. The role of tyrosine kinase activity in endocytosis, compartmentation, and down-regulation of the epidermal growth factor receptor. J Biol Chem, 266: 11083-11094, 1991.

91. Wikstrand, C. J., Hale, L. P., Batra, S. K., Hill, M.L., Humphrey, P. A., Kurpad, S. N., McLendon, R. E., Moscatello, D., Pegram, C. N., and Reist CJ, e. a. Monoclonal antibodies against EGFRvIII are tumor specific and react with breast and lung carcinomas and malignant gliomas. Cancer Res, 55: 3140-3148, 1995.

92. Moscatello, D. K., Holgado-Madruga, M., Godwin, A. K., Ramirez, G., Gunn, G., Zoltick, P. W., Biegel, J. A., Hayes, R. L., and Wong, A. J. Frequent expression of a mutant epidermal growth factor receptor in multiple human tumors. Cancer Res, 55: 5536-5539, 1995.

93. Sugawa, N., Ekstrand, A. J., James, C. D., and Collins, V. P. Identical splicing of aberrant epidermal growth factor receptor transcripts from amplified rearranged genes in human glioblastomas. Proc Natl Acad Sci USA, 87: 8602-8606, 1990.

94. Gilmore, T., DeClue, J. E., and Martin, G. S. Protein phosphorylation at tyrosine is induced by the v-erbB gene product in vivo and in vitro. Cell, 40: 609-618, 1985.

95. Kris, R. M., Lax, I., Gullick, W., Waterfield, M. D., Ullrich, A., Fridkin, M., and Schlessinger, J. Antibodies against a synthetic peptide as a probe for the kinase activity of the avian EGF receptor and v-erbB protein. Cell, 40: 619-625, 1985.

96. Coussens, L., Yang-Feng, T. L., Liao, Y. C., Chen, E., Gray, A., McGrath, J., Seeburg, P. H., Libermann, T. A., Schlessinger, J., and Francke, U. e. a. Tyrosine kinase receptor with extensive homology to EGF receptor shares chromosomal location with neu oncogene. Science, 230: 1132-1139, 1985.

97. Karunagaran, D., Tzahar, E., Beerli, R. R., Chen, X., Graus-Porta, D., Ratzkin, B. J., Seger, R., Hynes, N. E., and Yarden, Y. ErbB-2 is a common auxiliary subunit of NDF and EGF receptors: implications for breast cancer. EMBO J, 15: 254-264, 1996.

98. Ben-Levy, R., Peles, E., Goldman-Michael, R., and Yarden, Y. An oncogenic point mutation confers high affinity ligand binding to the neu receptor. Implications for the generation of site heterogeneity. J Biol Chem, 267: 17304-17313, 1992.

99. Stancovski, I., Peles, E., Ben Levy, R., Lemprecht, R., Kelman, Z., Goldman-Michael, R., Hurwitz, E., Bacus, S., Sela, M., and Yarden, Y. Signal transduction by the neu/erbB-2 receptor: a potential target for anti-tumor therapy. J Steroid Biochem Mol Biol, 43: 95-103, 1992.

100. Sternberg, M. J. and Gullick, W. J. Neu receptor dimerization. Nature, 339: 587, 1989.

101. Cao, H., Decker, S., and Stern, D. F. TPA inhibits the tyrosine kinase activity of the neu protein in vivo and in vitro. Oncogene, 6: 705-711, 1991.

102. Eccles, S. A., Modjtahedi, H., Box, G., Court, W., Sandle, J., and Dean, C. J. Significance of the c-erbB family of receptor tyrosine kinases in metastatic cancer and their potential as targets for immunotherapy. Invasion Metastasis, 14: 337-348., 1994-95.

103. Jardines, L., Weiss, M., Fowble, B., and Greene, M. neu(c-erbB-2/HER2) and the epidermal growth factor receptor (EGFR) in breast cancer. Pathobiology, 61: 268-282, 1993.

104. Kraus, M. H., Issing, W., Miki, T., Popescu, N. C., and Aaronson, S. A. Isolation and characterization of ERBB3, a third member of the ERBB/epidermal growth factor receptor family: evidence for overexpression in a subset of human mammary tumors. Proc Natl Acad Sci USA, 86: 9193-9197, 1989.

105. Plowman, G. D., Culouscou, J. M., Whitney, G. S., Green, J. M., Carlton, G. W., Foy, L., Neubauer, M. G., and Shoyab, M. Ligand-specific activation of

HER4/p180erbB4, a fourth member of the epidermal growth factor receptor family. Proc Natl Acad Sci USA, 90: 1746-1750, 1993.

106.   Gullick, W. J. The c-erbB3/HER3 receptor in human cancer. Cancer Surv, 27: 339-349, 1996.

107.   Carraway, K. L., Sliwkowski, M. X., Akita, R., Platko, J. V., Guy, P. M., Nuijens, A., Diamonti, A. J., Vandlen, R. L., Cantley, L. C., and Cerione, R. A. The erbB3 gene product is a receptor for heregulin. J Biol Chem, 269: 14303-14306, 1994.

108.   Soltoff, S. P., Carraway, K. L., Prigent, S. A., Gullick, W. G., and Cantley, L. C. ErbB3 is involved in activation of phosphatidylinositol 3-kinase by epidermal growth factor. Mol Cell Biol, 14: 3550-3558, 1994.

109.   Carroll, S. L., Miller, M. L., Frohnert, P. W., Kim, S. S., and Corbett, J. A. Expression of neuregulins and their putative receptors, ErbB2 and ErbB3, is induced during Wallerian degeneration. J Neurosci, 17: 1642-1659, 1997.

110.   Marchionni, M. A., Kirk, C. J., Isaacs, I. J., Hoban, C. J., Mahanthappa, N. K., Anton, E. S., Chen, C., Wason, F., Lawson, D., Hamers, F. P., Canoll, P. D., Reynolds, R., Cannella, B., Meun, D., Holt, W. F., Matthew, W. D., Chen, L. E., Gispen, W. H., Raine, C. S., Salzer, J. L., and Gwynne, D. I. Neuregulins as potential drugs for neurological disorders. Cold Spring Harb Symp Quant Biol, 61: 459-472, 1996.

111.   Riethmacher, D., Sonnenberg-Riethmacher, E., Brinkmann, V., Yamaai, T., Lewin, G. R., and Birchmeier, C. Severe neuropathies in mice with targeted mutations in the ErbB3 receptor. Nature, 389: 725-730, 1997.

112.   Bobrow, L. G., Millis, R. R., Happerfield, L. C., and Gullick, W. J. c-ErbB3 protein expression in ductal carcinoma in situ of the breast. Eur J Cancer, 33: 1846-1850, 1997.

113.   Zhang, K., Sun, J., Liu, N., Wen, D., Chang, D., Thomason, A., and Yoshinaga, S. K. Transformation of NIH 3T3 cells by HER3 or HER4 receptors requires the presence of HER1 or HER2. J Biol Chem, 271: 3884-3890, 1996.

114.   Elenius, K., Corfas, G., Paul, S., Choi, C. J., Rio, C., Plowman, G. D., and Klagsbrun, M. A novel juxtamembrane domain isoform of HER4/ErbB4. Isoform-

specific tissue distribution and differential processing in response to phorbol ester. J Biol Chem, 272: 26761-26768, 1997.

115. Vecchi, M., Baulida, J., and Carpenter, G. Selective cleavage of the heregulin receptor ErbB4 by protein kinase C activation. J Biol Chem, 271: 18989-18995, 1996.

116. Gassmann, M., Casagranda, F., Orioli, D., Simon, H., Lai, C., Klein, R., and Lemke, G. Aberrant neural and cardiac development in mice lacking the ErbB4 neuregulin receptor. Nature, 378: 390-394, 1995.

117. Srinivasan, R., Poulsom, R., Hurst, H. C., and Gullick, W. J. Expression of the c-ErbB4/HER4 protein and mRNA in normal human fetal and adult tissues and in a survey of nine solid tumour types. J Pathol, 185: 236-245, 1998.

118. Pelicci, G., Lanfrancone, L., Grignani, F., McGlade, J., Cavallo, F., Forni, G., Nicoletti, I., Pawson, T., and Pelicci, P. G. A novel transforming protein (SHC) with an SH2 domain is implicated in mitogenic signal transduction. Cell, 70: 93-104, 1992.

119. Migliaccio, E., Mele, S., Salcini, A. E., Pelicci, G., Lai, K. M., Superti-Furga, G., Pawson, T., Di Fiore, P. P., Lanfrancone, L., and Pelicci, P. G. Opposite effects of the p52shc/p46shc and p66shc splicing isoforms on the EGF receptor-MAP kinase-fos signalling pathway. EMBO J, 16: 706-716, 1997.

120. Bonfini, L., Migliaccio, E., Pelicci, G., Lanfrancone, L., and Pelicci, P. G. Not all Shc's roads lead to Ras Trends Biochem Sci, 21: 257-261, 1996.

121. Marshall, M. S. Ras target proteins in eukaryotic cells. FASEB J, 9. 1311 1318, 1995.

122. Salcini, A. E., McGlade, J., Pelicci, G., Nicoletti, I., Pawson, T., and Pelicci, P. G. Formation of Shc–Grb2 complexes is necessary to induce neoplastic transformation by overexpression of Shc proteins. Oncogene, 9: 2827-2836, 1994.

123. Gotoh, N., Toyoda, M., and Shibuya, M. Tyrosine phosphorylation sites at amino acids 239 and 240 of Shc are involved in epidermal growth factor-induced mitogenic signalling that is distinct from Ras/mitogen-activated protein kinase activation. Mol Cell Biol, 17: 1824-1831, 1997.

124.   Lanfrancone, L., Pelicci, G., Brizzi, M. F., Aronica, M. G., Casciari, C., Giuli, S., Pegoraro, L., Pawson, T., Pelicci, P. G., and Arouica, M. G. Overexpression of Shc proteins potentiates the proliferative response to the granulocyte-macrophage colony-stimulating factor and recruitment of Grb2/SoS and Grb2/p140 complexes to the beta receptor subunit. Oncogene, 10: 907-917, 1995.

125.   Lai, K. M. and Pawson, T. The ShcA phosphotyrosine docking protein sensitizes cardiovascular signalling in the mouse embryo. Genes Dev, 14: 1132-1145, 2000.

126.   Migliaccio, E., Giorgio, M., Mele, S., Pelicci, G., Reboldi, P., Pandolfi, P. P., Lanfrancone, L., and Pelicci, P. G. The p66shc adaptor protein controls oxidative stress response and life span in mammals. Nature, 402: 309-313, 1999.

127.   Okada, S., Kao, A. W., Ceresa, B. P., Blaikie, P., Margolis, B., and Pessin, J. E. The 66-kDa Shc isoform is a negative regulator of the epidermal growth factor-stimulated mitogen-activated protein kinase pathway. J Biol Chem, 272: 28042-28049, 1997.

128.   Pelicci, G., Dente, L., De Giuseppe, A., Verducci-Galletti, B., Giuli, S., Mele, S., Vetriani, C., Giorgio, M., Pandolfi, P. P., and Cesareni, G. e. a. A family of Shc related proteins with conserved PTB, CH1 and SH2 regions. Oncogene, 13: 633-641, 1996.

129.   Nakamura, T., Muraoka, S., Sanokawa, R., and Mori, N. N-Shc and Sck, two neuronally expressed Shc adapter homologs. Their differential regional expression in the brain and roles in neurotrophin and Src signalling. J Biol Chem, 273: 6960-6967, 1998.

130.   Yang, J., Cron, P., Good, V. M., Thompson, V., Hemmings, B. A., and Barford, D. Crystal structure of an activated Akt/protein kinase B ternary complex with GSK3-peptide and AMP-PNP. Nat Struct Biol, 9: 940-944, 2002.

131.   Thomas, C. C., Deak, M., Alessi, D. R., and van Aalten, D. M. High-resolution structure of the pleckstrin homology domain of protein kinase b/akt bound to phosphatidylinositol (3,4,5)-trisphosphate. Curr Biol, 12: 1256-1262, 2002.

132.	Maehama, T. and Dixon, J. E. PTEN: a tumour suppressor that functions as a phospholipid phosphatase. Trends Cell Biol, 9,: 125-128, 1999.

133.	Stambolic, V., Suzuki, A., de la Pompa, J. L., Brothers, G. M., Mirtsos, C., Sasaki, T., Ruland, J., Penninger, J. M., Siderovski, D. P., and Mak, T. W. Negative regulation of PKB/Akt-dependent cell survival by the tumor suppressor PTEN. Cell, 95: 29-39, 1998.

134.	Damen, J. E., Liu, L., Rosten, P., Humphries, R. K., Jejerson, A. B., Majerus, P. W., and Krystal, G. The 145-kDa protein induced to associate with Shc by multiple cytokines is an inositol tetraphosphate and phosphatidylinositol 3,4,5-triphosphate 5-phosphatase. Proc Natl Acad Sci USA, 93: 1689-1693, 1996.

135.	Pearson, R. B., Dennis, P. B., Han, J. W., Williamson, N. A., Kozma, S. C., Wettenhall, R. E., and Thomas, G. The principal target of rapamycin-induced p70s6k inactivation is a novel phosphorylation site within a conserved hydrophobic domain. EMBO J, 14: 5279-5287, 1995.

136.	Balendran, A., Biondi, R. M., Cheung, P. C., Casamayor, A., Deak, M., and Alessi, D. R. A 3-phosphoinositide-dependent protein kinase-1 (PDK1) docking site is required for the phosphorylation of protein kinase Czeta (PKCzeta) and PKC-related kinase 2 by PDK1. J Biol Chem, 275: 20806-20813, 2000.

137.	Frodin, M., Antal, T. L., Dummler, B. A., Jensen, C. J., Deak, M., Gammeltoft, S., and Biondi, R. M. A phosphoserine/threonine-binding pocket in AGC kinases and PDK1 mediates activation by hydrophobic motif phosphorylation. EMBO J, 21: 5396-5407, 2002.

138.	Biondi, R. M., Kieloch, A., Currie, R. A., Deak, M., and Alessi, D. R. The PIF-binding pocket in PDK1 is essential for activation of S6K and SGK, but not PKB. EMBO J, 20: 4380-4390, 2001.

139.	Yang, J., Cron, P., Thompson, V., Good, V. M., Hess, D., Hemmings, B. A., and Barford, D. Molecular mechanism for the regulation of protein kinase B/Akt by hydrophobic motif phosphorylation. Mol Cell, 9: 1227-1240, 2002.

140.	Batkin, M., Schvartz, I., and Shaltiel, S. Snapping of the carboxyl terminal tail of the catalytic subunit of PKA onto its core: characterization of the sites by mutagenesis. Biochemistry, 39: 5366-5373, 2000.

141.   Alessi, D. R., Andjelkovic, M., Caudwell, B., Cron, P., Morrice, N., Cohen, P., and Hemmings, B. A. Mechanism of activation of protein kinase B by insulin and IGF-1. EMBO J, 15: 6541-6551, 1996.

142.   Alessi, D. R., James, S. R., Downes, C. P., Holmes, A. B., Gaffney, P. R., Reese, C. B., and Cohen, P. Characterization of a 3-phosphoinositide-dependent protein kinase which phosphorylates and activates protein kinase Balpha. Curr Biol, 7: 261-269, 1997.

143.   Stephens, L., Anderson, K., Stokoe, D., Erdjument-Bromage, H., Painter, G. F., Holmes, A. B., Gaɪney, P. R., Reese, C. B., McCormick, F., Tempst, P., Coadwell, J., and Hawkins, P. T. Protein kinase B kinases that mediate phosphatidylinositol 3,4,5-trisphosphate-dependent activation of protein kinase B. Science, 279: 710-714, 1998.

144.   Williams, M. R., Arthur, J. S., Balendran, A., Van Der, K. J., Poli, V., Cohen, P., and Alessi, D. R. The role of 3-phosphoinositide-dependent protein kinase 1 in activating AGC kinases defined in embryonic stem cells. Curr Biol, 10: 439-448, 2000.

145.   Flynn, P., Wongdagger, M., Zavar, M., Dean, N. M., and Stokoe, D. Inhibition of PDK-1 activity causes a reduction in cell proliferation and survival. Curr Biol, 10: 1439-1442, 2000.

146.   Toker, A. and Newton, A. C. Akt/protein kinase B is regulated by autophosphorylation at the hypothetical PDK-2 site. J Biol Chem, 275: 8271-8274, 2000.

147.   Delcommenne, M., Tan, C., Gray, V., Ruel, L., Woodgett, J., and Dedhar, S. Phosphoinositide-3-OH kinase-dependent regulation of glycogen synthase kinase 3 and protein kinase B/AKT by the integrin-linked kinase. Proc Natl Acad Sci USA, 95: 11211-11216, 1998.

148.   Persad, S., Attwell, S., Gray, V., Mawji, N., Deng, J. T., Leung, D., Yan, J., Sanghera, J., Walsh, M. P., and Dedhar, S. Regulation of protein kinase B/Akt-serine 473 phosphorylation by integrin-linked kinase: critical roles for kinase activity and amino acids arginine 211 and serine 343. J Biol Chem, 276: 27462-27469, 2001.

149. Chen, R., Kim, O., Yang, J., Sato, K., Eisenmann, K. M., McCarthy, J., Chen, H., and Qiu, Y. Regulation of Akt/PKB activation by tyrosine phosphorylation. J Biol Chem, 276: 31858-31862, 2001.

150. Jiang, T. and Qiu, Y. Interaction between Src and a C-terminal proline-rich motif of Akt is required for Akt activation. J Biol Chem, 278: 15789-15793, 2003.

151. Conus, N. M., Hannan, K. M., Cristiano, B. E., Hemmings, B. A., and Pearson, R. B. Direct identification of tyrosine 474 as a regulatory phosphorylation site for the Akt protein kinase. J Biol Chem, 277: 38021-38028, 2002.

152. Dudek, H. e. a. Regulation of neuronal survival by the serinethreonine protein kinase Akt. Science, 275: 661–665, 1997.

153. Li, J. e. a. The PTEN/MMAC1 tumor suppressor induces cell death that is rescued by the AKT/protein kinase B oncogene. Cancer Res, 58: 5667–5672, 1998.

154. Datta, S. R. e. a. Akt phosphorylation of BAD couples survival signals to the cell-intrinsic death machinery. Cell, 91: 231–241, 1997.

155. Cardone, M. H. e. a. Regulation of cell death protease caspase-9 by phosphorylation. Science, 282: 1318–1321, 1998.

156. Brunet, A. e. a. Akt promotes cell survival by phosphorylating and inhibiting a Forkhead transcription factor. Cell, 96: 857–868, 1999.

157. Romashkova, J. A. and Makarov, S. S. NF-kB is a target of AKT in anti-apoptotic PDGF signalling. Nature, 401: 86–90, 1999.

158. Kane, L. P., Shapiro, V. S., Stokoe, D., and Weiss, A. Induction of NF-kB by the Akt/PKB kinase. Curr Biol, 9: 601–604, 1999.

159. Mayo, L. D. and Donner, D. B. A phosphatidylinositol kinase/Akt pathway promotes translocation of Mdm2
from the cytoplasm to the nucleus. Proc Natl Acad Sci USA, 98,: 11598–11603, 2001.

160. Zhou, B. P. e. a. HER-2/neu induces p53 ubiquitination via Akt-mediated MDM2 phosphorylation. Nature Cell Biol, 3: 973–982, 2001.

161.	Diehl, J. A., Cheng, M., Roussel, M. F., and Sherr, C. J. Glycogen synthase kinase-3b regulates cyclin D1 proteolysis and subcellular localization. Genes Dev, 12: 3499–3511, 1998.

162.	Graff, J. R. e. a. Increased AKT activity contributes to prostate cancer progression by dramatically accelerating prostate tumor growth and diminishing p27Kip1 expression. J Biol Chem, 275: 24500–24505, 2000.

163.	Vemuri, G. S. and Rittenhouse, S. E. Wortmannin inhibits serum-induced activation of phosphoinositide 3-kinase and proliferation of CHRF-288 cells. Biochem Biophys Res Commun, 202: 1619–1623, 1994.

164.	Castoria, G. e. a. PI3-kinase in concert with Src promotes the S-phase entry of oestradiol-stimulated MCF-7 cells. EMBO J, 20: 6050–6059, 2001.

165.	Nave, B. T., Ouwens, M., Withers, D. J., Alessi, D. R., and Shepherd, P. R. Mammalian target of rapamycin is a direct target for protein kinase B: identification of a convergence point for opposing effects of insulin and amino-acid deficiency on protein translation. Biochem J, 344: 427–431, 1999.

166.	Hunter, T. and Cooper, J. A. Protein-tyrosine kinases. Annu Rev Biochem, 54: 897–930, 1985.

167.	Fantl, W. J., Johnson, D. E., and Williams, L. T. Signalling by receptor tyrosine kinases. Annu Rev Biochem, 62: 453–481, 1993.

168.	Wang, J. Y. L. Isolation of antibodies for phosphotyrosine by immunization with a v-abl oncogene-encoded protein. Mol Cell Biol, 5: 3640–3643, 1985.

169.	Glenney, J. R., Jr., Zokas, L., and Kamps, M. P. Monoclonal antibodies to phosphotyrosine. J Immunol Methods, 109, 1988.

170.	Slamon, D. J., Clark, G. M., Wong, S. G., Levin, W. J., Ullrich, A., and McGuire, W. L. Human breast cancer: correlation of relapse and survival with amplification of the HER-2/neu oncogene. Science, 235: 177-182, 1987.

171.	Eppenberger-Castori, S., Kueng, W., Benz, C., Caduff, R., Varga, Z., Bannwart, F., Fink, D., Dieterich, H., Hohl, M., Muller, H., Paris, K., Schoumacher, F., and Eppenberger, U. Prognostic and predictive significance of ErbB-2 breast tumor levels measured by enzyme immunoassay. J Clin Oncol, 19: 645-656, 2001.

172.   Konecny, G., Pauletti, G., Pegram, M., Untch, M., Dandekar, S., Aguilar, Z., Wilson, C., Rong, H. M., Bauerfeind, I., Felber, M., Wang, H. J., Beryt, M., Seshadri, R., Hepp, H., and Slamon, D. J. Quantitative association between HER-2/neu and steroid hormone receptors in hormone receptor-positive primary breast cancer. J Natl Cancer Inst, 95: 142-153, 2003.

173.   Hayes, D. F. and Thor, A. D. c-erbB-2 in breast cancer: development of a clinically useful marker. Semin Oncol, 29: 231-245, 2002.

174.   Wright, C., Nicholson, S., Angus, B., Sainsbury, J. R., Farndon, J., Cairns, J., Harris, A. L., and Horne, C. H. Relationship between c-erbB-2 protein product expression and response to endocrine therapy in advanced breast cancer. Br J Cancer, 65: 118-121, 1992.

175.   Nicholson, R. I., McClelland, R. A., Finlay, P., Eaton, C. L., Gullick, W. J., Dixon, A. R., Robertson, J. F., Ellis, I. O., and Blamey, R. W. Relationship between EGF-R, c-erbB-2 protein expression and Ki67 immunostaining in breast cancer and hormone sensitivity. Eur J Cancer, 29A: 1018-1023, 1993.

176.   Dowsett, M., Harper-Wynne, C., Boeddinghaus, I., Salter, J., Hills, M., Dixon, M., Ebbs, S., Gui, G., Sacks, N., and Smith, I. HER-2 amplification impedes the antiproliferative effects of hormone therapy in estrogen receptor-positive primary breast cancer. Cancer Res, 61: 8452-8458, 2001.

177.   Schechter, A. L., Hung, M. C., Vaidyanathan, L., Weinberg, R. A., Yang-Feng, T. L., Francke, U., Ullrich, A., and Coussens, L. The neu gene: an erbB-homologous gene distinct from and unlinked to the gene encoding the EGF receptor. Science, 229: 976-978, 1985.

178.   Coussens, L., Yang-Feng, T. L., Liao, Y. C., Chen, E., Gray, A., McGrath, J., Seeburg, P. H., Libermann, T. A., Schlessinger, J., Francke, U., and ullrich, A. Tyrosine kinase receptor with extensive homology to EGF receptor shares chromosomal location with neu oncogene. Science, 230: 1132-1139, 1985.

179.   Bargmann, C. I., Hung, M. C., and Weinberg, R. A. The neu oncogene encodes an epidermal growth factor receptor-related protein. Nature, 319: 226-230, 1986.

180.   Yamamoto, T., Ikawa, S., Akiyama, T., Semba, K., Nomura, N., Miyajima, N., Saito, T., and Toyoshima, K. Similarity of protein encoded by the human c-erb-B-2 gene to epidermal growth factor receptor. Nature, 319: 230-234, 1986.

181.   Hudziak, R. M., Schlessinger, J., and Ullrich, A. Increased expression of the putative growth factor receptor p185HER2 causes transformation and tumorigenesis of NIH 3T3 cells. Proc Natl Acad Sci USA, 84: 7159-7163, 1987.

182.   Di Fiore, P. P., Pierce, J. H., Kraus, M. H., Segatto, O., King, C. R., and Aaronson, S. A. erbB-2 is a potent oncogene when overexpressed in NIH/3T3 cells. Science, 237: 178-182, 1987.

183.   Stern, D. F., Kamps, M. P., and Cao, H. Oncogenic activation of p185neu stimulates tyrosine phosphorylation in vivo. Mol Cell Biol, 8: 3969-3973, 1988.

184.   Pierce, J. H., Arnstein, P., DiMarco, E., Artrip, J., Kraus, M. H., Lonardo, F., Di Fiore, P. P., and Aaronson, S. A. Oncogenic potential of erbB-2 in human mammary epithelial cells. Oncogene, 6: 1189-1194, 1991.

185.   Bargmann, C. I., Hung, M. C., and Weinberg, R. A. Multiple independent activations of the neu oncogene by a point mutation altering the transmembrane domain of p185. Cell, 45: 649-657, 1986.

186.   Bargmann, C. I. and Weinberg, R. A. Oncogenic activation of the neu-encoded receptor protein by point mutation and deletion. EMBO J, 7: 2043-2052, 1988.

187.   Weiner, D. B., Liu, J., Cohen, J. A., Williams, W. V., and Greene, M. I. A point mutation in the neu oncogene mimics ligand induction of receptor aggregation. Nature, 339: 230-231, 1989.

188.   Siegel, P. M., Dankort, D. L., Hardy, W. R., and Muller, W. J. Novel activating mutations in the neu proto-oncogene involved in induction of mammary tumors. Mol Cell Biol, 14: 7068-7077, 1994.

189.   Christianson, T. A., Doherty, J. K., Lin, Y. J., Ramsey, E. E., Holmes, R., Keenan, E. J., and Clinton, G. M. NH2-terminally truncated HER-2/neu protein: relationship with shedding of the extracellular domain and with prognostic factors in breast cancer. Cancer Res, 58: 5123-5129, 1998.

190. Molina, M. A., Codony-Servat, J., Albanell, J., Rojo, F., Arribas, J., and Baselga, J. Trastuzumab (herceptin), a humanized anti-Her2 receptor monoclonal antibody, inhibits basal and activated Her2 ectodomain cleavage in breast cancer cells. Cancer Res, 61: 4744-4749, 2001.

191. Reese, D. M., Small, E. J., Magrane, G., Waldman, F. M., Chew, K., and Sudilovsky, D. HER2 protein expression and gene amplification in androgen-independent prostate cancer. Am J Clin Pathol, 116: 234-239, 2001.

192. Cox, G., Vyberg, M., Melgaard, B., Askaa, J., Oster, A., and O'Byrne, K. J. Herceptest: HER2 expression and gene amplification in non-small cell lung cancer. Int J Cancer, 92: 480-483, 2001.

193. Khan, A. J., King, B. L., Smith, B. D., Smith, G. L., DiGiovanna, M. P., Carter, D., and Haffty, B. G. Characterization of the HER-2/neu oncogene by immunohistochemical and fluorescence in situ hybridization analysis in oral and oropharyngeal squamous cell carcinoma. Clin Cancer Res, 8: 540-548, 2002.

194. DiGiovanna, M. P. and Stern, D. F. Activation state-specific monoclonal antibody detects tyrosine phosphorylated p185neu/erbB-2 in a subset of human breast tumors overexpressing this receptor. Cancer Res, 55: 1946-1955, 1995.

195. DiGiovanna, M. P., Carter, D., Flynn, S. D., and Stern, D. F. Functional assay for HER-2/neu demonstrates active signalling in a minority of HER-2/neu-overexpressing invasive human breast tumours. Br J Cancer, 74: 802-806, 1996.

196. Thor, A. D., Liu, S., Edgerton, S., Moore, D., 2nd, Kasowitz, K. M., Benz, C. C., Stern, D. F., and DiGiovanna, M. P. Activation (tyrosine phosphorylation) of ErbB-2 (HER-2/neu): a study of incidence and correlation with outcome in breast cancer. J Clin Oncol, 18: 3230-3239, 2000.

197. DiGiovanna, M. P., Chu, P., Davison, T. L., Howe, C. L., Carter, D., Claus, E. B., and Stern, D. F. Active signaling by HER-2/neu in a subpopulation of HER-2/neu-overexpressing ductal carcinoma in situ: clinicopathological correlates. Cancer Res, 62: 6667-6673, 2002.

198. Hudelist, G., Kostler, W. J., Attems, J., Czerwenka, K., Muller, R., Manavi, M., Steger, G. G., Kubista, E., Zielinski, C. C., and Singer, C. F. Her-2/neu-triggered intracellular tyrosine kinase activation: in vivo relevance of ligand-

independent activation mechanisms and impact upon the efficacy of trastuzumab-based treatment. Br J Cancer, 89: 983-991, 2003.

199. Knowlden, J. M., Gee, J. M., Seery, L. T., Farrow, L., Gullick, W. J., Ellis, I. O., Blamey, R. W., Robertson, J. F., and Nicholson, R. I. c-erbB3 and c-erbB4 expression is a feature of the endocrine responsive phenotype in clinical breast cancer. Oncogene, 17: 1949-1957, 1998.

200. Pawlowski, V., Revillion, F., Hebbar, M., Hornez, L., and Peyrat, J. P. Prognostic value of the type I growth factor receptors in a large series of human primary breast cancers quantified with a real-time reverse transcription-polymerase chain reaction assay. Clin Cancer Res, 6: 4217-4225, 2000.

201. Suo, Z., Risberg, B., Kalsson, M. G., Willman, K., Tierens, A., Skovlund, E., and Nesland, J. M. EGFR family expression in breast carcinomas. c-erbB-2 and c-ErbB4 receptors have different effects on survival. J Pathol, 196: 17-25, 2002.

202. Holbro, T., Beerli, R. R., Maurer, F., Koziczak, M., Barbas, C. F., and Hynes, N. E. The ErbB2/ErbB3 heterodimer functions as an oncogenic unit: ErbB2 requires ErbB3 to drive breast tumor cell proliferation. Proc Natl Acad Sci USA, 100: 8933-8938, 2003.

203. Brandt, B. H., Roetger, A., Dittmar, T., Nikolai, G., Seeling, M., Merschjann, A., Nofer, J. R., Dehmer-Moller, G., Junker, R., Assmann, G., and Zaenker, K. S. c-erbB-2/EGFR as dominant heterodimerization partners determine a motogenic phenotype in human breast cancer cells. Faseb J, 13: 1939-1949, 1999.

204. Dittmar, T., Husemann, A., Schewe, Y., Nofer, J. R., Niggemann, B., Zanker, K. S., and Brandt, B. H. Induction of cancer cell migration by epidermal growth factor is initiated by specific phosphorylation of tyrosine 1248 of c-erbB-2 receptor via EGFR. Faseb J, 16: 1823-1825, 2002.

205. Vivanco, I. and Sawyers, C. L. The phosphatidylinositol 3-Kinase AKT pathway in human cancer. Nat Rev Cancer, 2: 489-501, 2002.

206. Hanada, M., Feng, J., and Hemmings, B. A. Structure, regulation and function of PKB/AKT--a major therapeutic target. Biochim Biophys Acta, 1697: 3-16, 2004.

207. Staal, S. P. Molecular cloning of the akt oncogene and its human homologues AKT1 and AKT2: amplification of AKT1 in a primary human gastric adenocarcinoma. Proc Natl Acad Sci USA, 84: 5034-5037, 1987.

208. Franke, T. F., Kaplan, D. R., and Cantley, L. C. PI3K: downstream AKTion blocks apoptosis. Cell, 88: 435-437, 1997.

209. Bowers, D. C., Fan, S., Walter, K. A., Abounader, R., Williams, J. A., Rosen, E. M., and Laterra, J. Scatter factor/hepatocyte growth factor protects against cytotoxic death in human glioblastoma via phosphatidylinositol 3-kinase- and AKT-dependent pathways. Cancer Res, 60: 4277-4283, 2000.

210. Hii, C. S., Moghadammi, N., Dunbar, A., and Ferrante, A. Activation of the phosphatidylinositol 3-kinase-Akt/protein kinase B signaling pathway in arachidonic acid-stimulated human myeloid and endothelial cells: involvement of the ErbB receptor family. J Biol Chem, 276: 27246-27255, 2001.

211. Chen, Y., Li, X., Eswarakumar, V. P., Seger, R., and Lonai, P. Fibroblast growth factor (FGF) signaling through PI 3-kinase and Akt/PKB is required for embryoid body differentiation. Oncogene, 19: 3750-3756, 2000.

212. Duan, C., Liimatta, M. B., and Bottum, O. L. Insulin-like growth factor (IGF)-I regulates IGF-binding protein-5 gene expression through the phosphatidylinositol 3-kinase, protein kinase B/Akt, and p70 S6 kinase signaling pathway. J Biol Chem, 274: 37147-37153, 1999.

213. Franke, T. F., Yang, S. I., Chan, T. O., Datta, K., Kazlauskas, A., Morrison, D. K., Kaplan, D. R., and Tsichlis, P. N. The protein kinase encoded by the Akt proto-oncogene is a target of the PDGF-activated phosphatidylinositol 3-kinase. Cell, 81: 727-736, 1995.

214. Cantley, L. C. and Neel, B. G. New insights into tumor suppression: PTEN suppresses tumor formation by restraining the phosphoinositide 3-kinase/AKT pathway. Proc Natl Acad Sci USA, 96: 4240-4245, 1999.

215. Miwa, W., Yasuda, J., Murakami, Y., Yashima, K., Sugano, K., Sekine, T., Kono, A., Egawa, S., Yamaguchi, K., Hayashizaki, Y., and Sekiya, T. Isolation of DNA sequences amplified at chromosome 19q13.1-q13.2 including the AKT2 locus in human pancreatic cancer. Biochem Biophys Res Commun, 225: 968-974, 1996.

216. Cheng, J. Q., Godwin, A. K., Bellacosa, A., Taguchi, T., Franke, T. F., Hamilton, T. C., Tsichlis, P. N., and Testa, J. R. AKT2, a putative oncogene encoding a member of a subfamily of protein-serine/threonine kinases, is amplified in human ovarian carcinomas. Proc Natl Acad Sci USA, 89: 9267-9271, 1992.

217. Bellacosa, A., de Feo, D., Godwin, A. K., Bell, D. W., Cheng, J. Q., Altomare, D. A., Wan, M., Dubeau, L., Scambia, G., Masciullo, V., and al., e. Molecular alterations of the AKT2 oncogene in ovarian and breast carcinomas. Int J Cancer, 64: 280-285, 1995.

218. Roy, H. K., Olusola, B. F., Clemens, D. L., Karolski, W. J., Ratashak, A., Lynch, H. T., and Smyrk, T. C. AKT proto-oncogene overexpression is an early event during sporadic colon carcinogenesis. Carcinogenesis, 23: 201-205, 2002.

219. Ruggeri, B. A., Huang, L., Wood, M., Cheng, J. Q., and Testa, J. R. Amplification and overexpression of the AKT2 oncogene in a subset of human pancreatic ductal adenocarcinomas. Mol Carcinog, 21: 81-86, 1998.

220. Bacus, S. S., Altomare, D. A., Lyass, L., Chin, D. M., Farrell, M. P., Gurova, K., Gudkov, A., and Testa, J. R. AKT2 is frequently upregulated in HER-2/neu-positive breast cancers and may contribute to tumor aggressiveness by enhancing cell survival. Oncogene, 21: 3532-3540, 2002.

221. Perez-Tenorio, G. and Stal, O. Activation of AKT/PKB in breast cancer predicts a worse outcome among endocrine treated patients. Br J Cancer, 86: 540-545, 2002.

222. Stal, O., Perez-Tenorio, G., Akerberg, L., Olsson, B., Nordenskjold, B., Skoog, L., and Rutqvist, L. E. Akt kinases in breast cancer and the results of adjuvant therapy. Breast Cancer Res, 5: R37-44, 2003.

223. Schmitz, K. J., Otterbach, F., Callies, R., Levkau, B., Holscher, M., Hoffmann, O., Grabellus, F., Kimmig, R., Schmid, K. W., and Baba, H. A. Prognostic relevance of activated Akt kinase in node-negative breast cancer: a clinicopathological study of 99 cases. Mod Pathol, 17: 15-21, 2004.

224. Malik, S. N., Brattain, M., Ghosh, P. M., Troyer, D. A., Prihoda, T., Bedolla, R., and Kreisberg, J. I. Immunohistochemical demonstration of phospho-Akt in high Gleason grade prostate cancer. Clin Cancer Res, 8: 1168-1171, 2002.

225.   Liao, Y., Grobholz, R., Abel, U., Trojan, L., Michel, M. S., Angel, P., and Mayer, D. Increase of AKT/PKB expression correlates with gleason pattern in human prostate cancer. Int J Cancer, 107: 676-680, 2003.

226.   Yamamoto, S., Tomita, Y., Hoshida, Y., Morooka, T., Nagano, H., Dono, K., Umeshita, K., Sakon, M., Ishikawa, O., Ohigashi, H., Nakamori, S., Monden, M., and Aozasa, K. Prognostic significance of activated Akt expression in pancreatic ductal adenocarcinoma. Clin Cancer Res, 10: 2846-2850, 2004.

227.   Ross, J. S. and Fletcher, J. A. HER-2/neu (c-erb-B2) gene and protein in breast cancer. Am J Clin Pathol, 112: S53-67, 1999.

228.   Hellyer, N. J., Kim, M. S., and Koland, J. G. Heregulin-dependent activation of phosphoinositide 3-kinase and Akt via the ErbB2/ErbB3 co-receptor. J Biol Chem, 276: 42153-42161, 2001.

229.   Murga, C., Laguinge, L., Wetzker, R., Cuadrado, A., and Gutkind, J. S. Activation of Akt/protein kinase B by G protein-coupled receptors. A role for alpha and beta gamma subunits of heterotrimeric G proteins acting through phosphatidylinositol-3-OH kinase gamma. J Biol Chem, 273: 19080-19085, 1998.

230.   Mills, G. B., Lu, Y., Fang, X., Wang, H., Eder, A., Mao, M., Swaby, R., Cheng, K. W., Stokoe, D., Siminovitch, K., Jaffe, R., and Gray, J. The role of genetic abnormalities of PTEN and the phosphatidylinositol 3-kinase pathway in breast and ovarian tumorigenesis, prognosis, and therapy. Semin Oncol, 28: 125-141, 2001.

231.   Muise-Helmericks, R. C., Grimes, H. L., Bellacosa, A., Malstrom, S. E., Tsichlis, P. N., and Rosen, N. Cyclin D expression is controlled post-transcriptionally via a phosphatidylinositol 3-kinase/Akt-dependent pathway. J Biol Chem, 273: 29864-29872, 1998.

232.   Graff, J. R., Konicek, B. W., McNulty, A. M., Wang, Z., Houck, K., Allen, S., Paul, J. D., Hbaiu, A., Goode, R. G., Sandusky, G. E., Vessella, R. L., and Neubauer, B. L. Increased AKT activity contributes to prostate cancer progression by dramatically accelerating prostate tumor growth and diminishing p27Kip1 expression. J Biol Chem, 275: 24500-24505, 2000.

233.   Slichenmyer, W. J. and Fry, D. W. Anticancer therapy targeting the erbB family of receptor tyrosine kinases. Semin Oncol, 28: 67-79, 2001.

234. Akiyama, T., Matsuda, S., Namba, Y., Saito, T., Toyoshima, K., and Yamamoto, T. The transforming potential of the c-erbB-2 protein is regulated by its autophosphorylation at the carboxyl-terminal domain. Mol Cell Biol, 11: 833-842, 1991.

235. Davis, R. J. and Czech, M. P. Platelet-derived growth factor mimics phorbol diester action on epidermal growth factor receptor phosphorylation threonine-654. Proc Natl Acad Sci USA, 82: 4080–4084, 1985.

236. Davis, R. J. and Czech, M. P. Tumor-promoting phorbol diesters cause the phosphorylation of epidermal growth factor receptors in normal human fibroblasts at threonine-654. Proc Natl Acad Sci USA, 82: 1974–1978, 1985.

237. Lin, C. R., Chen, W. S., Lazar, C. S., Carpenter, C. D., Gill, G. N., Evans, R. M., and Rosenfeld, M. G. Protein kinase C phosphorylation at Thr 654 of the unoccupied EGF receptor and EGF binding regulate functional receptor loss by independent mechanisms. Cell, 44: 839–848, 1986.

238. Countaway, J. L., Nairn, A. C., and Davis, R. J. Mechanism of desensitization of the epidermal growth factor receptor protein-tyrosine kinase. J Biol Chem, 267: 1129–1140, 1992.

239. Gotoh, N., Tojo, A., and Shibuya, M. A novel pathway from phosphorylation of tyrosine residues 239/240 of Shc, contributing to suppress apoptosis by IL-3. EMBO J, 15: 6197-6204, 1996.

240. Gotoh, N., Toyoda, M., and Shibuya, M. Tyrosine phosphorylation sites at amino acids 239 and 240 of Shc are involved in epidermal growth factor-induced mitogenic signaling that is distinct from Ras/mitogen-activated protein kinase activation. Mol Cell Biol, 17: 1824 -1831, 1997.

241. Velazquez, L., Gish, G. D., van Der Geer, P., Taylor, L., Shulman, J., and Pawson, T. The shc adaptor protein forms interdependent phosphotyrosine-mediated protein complexes in mast cells stimulated with interleukin 3. Blood, 96: 132-138, 2000.

242. Le, S., Connors, T. J., and Maroney, A. C. c-Jun N-terminal kinase specifically phosphorylates p66ShcA at serine 36 in response to ultraviolet irradiation. J Biol Chem, 276: 48332-48336.

243.   Huang, G. C., Ouyang, X., and Epstein, R. J. Proxy activation of ErbB2 by heterologous ligands suggests a heterotetrameric mechanism of receptor tyrosine kinase interaction. Biochem J, 331: 113–119, 1998.

244.   Ouyang, X., Gulliford, T., Zhang, H., Huang, G. C., and Epstein, R. J. Human cancer cells exhibit protein kinase C-dependent c-erbB-2 transmodulation which correlates with phosphatase sensitivity and kinase activity. J Biol Chem, 271: 21786–21792, 1996.

245.   Ouyang, X., Huang, G. C., Chantry, A., and Epstein, R. J. Adjacent carboxyterminal tyrosine phosphorylation events identify functionally distinct ErbB2 receptor subsets. Exp Cell Res, 241: 467–475, 1998.

246.   Gulliford, T., Huang, G. C., Ouyang, X., and Epstein, R. J. Reduced ability of transforming growth factor-alpha
induce hetero-oligomerization and downregulation
of EGFR suggests a mechanism of oncogenic synergy
with ErbB2. Oncogene, 15: 2219–2223, 1997.

247.   Ouyang, X., Gulliford, T., Huang, G., and Epstein, R. J. Transforming growth factor-alpha short-circuits downregulation of the epidermal growth factor receptor. J Cell Physiol, 179: 52–57, 1999.

248.   Ouyang, X., Gulliford, T., and Epstein, R. J. The duration of phorbol-inducible ErbB2 tyrosine dephosphorylation parallels that of receptor endocytosis rather than threonine-686 phosphorylation: Implications for the physiologic role of protein kinase C in growth factor receptor signalling. Carcinogenesis, 19, 1998.

249.   Epstein, R. J., Druker, B. J., Roberts, T. M., and Stiles, C. D. Synthetic phosphopeptide immunogens yield activationspecific antibodies to the c-erbB-2 receptor. Proc Natl Acad Sci USA, 89: 10435–10439, 1992.

250.   Olayioye, M. A., Neve, R. M., Lane, H. A., and Hynes, N. E. The ErbB signaling network: receptor heterodimerization in development and cancer. Embo J, 19: 3159-3167, 2000.

251.   Benz, C. C., Scott, G. K., Sarup, J. C., Johnson, R. M., Tripathy, D., Coronado, E., Shepard, H. M., and Osborne, C. K. Estrogen-dependent, tamoxifen-

resistant tumorigenic growth of MCF-7 cells transfected with HER2/neu. Breast Cancer Res Treat, 24: 85-95, 1993.

252. Liu, Y., el-Ashry, D., Chen, D., Ding, I. Y., and Kern, F. G. MCF-7 breast cancer cells overexpressing transfected c-erbB-2 have an in vitro growth advantage in estrogen-depleted conditions and reduced estrogen-dependence and tamoxifen-sensitivity in vivo. Breast Cancer Res Treat, 34: 97-117, 1995.

253. Pietras, R. J., Arboleda, J., Reese, D. M., Wongvipat, N., Pegram, M. D., Ramos, L., Gorman, C. M., Parker, M. G., Sliwkowski, M. X., and Slamon, D. J. HER-2 tyrosine kinase pathway targets estrogen receptor and promotes hormone-independent growth in human breast cancer cells. Oncogene, 10: 2435-2446, 1995.

254. Nicholson, R. I., Hutcheson, I. R., Harper, M. E., Knowlden, J. M., Barrow, D., McClelland, R. A., Jones, H. E., Wakeling, A. E., and Gee, J. M. Modulation of epidermal growth factor receptor in endocrine-resistant, oestrogen receptor-positive breast cancer. Endocr Relat Cancer, 8: 175-182, 2001.

255. Newby, J. C., Johnston, S. R., Smith, I. E., and Dowsett, M. Expression of epidermal growth factor receptor and c-erbB2 during the development of tamoxifen resistance in human breast cancer. Clin Cancer Res, 3: 1643-1651, 1997.

256. Witters, L. M., Kumar, R., Chinchilli, V. M., and Lipton, A. Enhanced anti-proliferative activity of the combination of tamoxifen plus HER-2-neu antibody. Breast Cancer Res Treat, 42: 1-5, 1997.

257. Kurokawa, H. and Arteaga, C. L. Inhibition of erbB receptor (HER) tyrosine kinases as a strategy to abrogate antiestrogen resistance in human breast cancer. Clin Cancer Res, 7: 4436s-4442s; discussion 4411s-4412s, 2001.

258. Nicholson, R. I., Hutcheson, I. R., Harper, M. E., Knowlden, J. M., Barrow, D., McClelland, R. A., Jones, H. E., Wakeling, A. E., and Gee, J. M. Modulation of epidermal growth factor receptor in endocrine-resistant, estrogen-receptor-positive breast cancer. Ann N Y Acad Sci, 963: 104-115, 2002.

259. Witters, L., Engle, L., and Lipton, A. Restoration of estrogen responsiveness by blocking the HER-2/neu pathway. Oncol Rep, 9: 1163-1166, 2002.

260.   Kurokawa, H. and Arteaga, C. L. ErbB (HER) receptors can abrogate antiestrogen action in human breast cancer by multiple signaling mechanisms. Clin Cancer Res, 9: 511S-515S, 2003.

261.   Kurokawa, H., Lenferink, A. E., Simpson, J. F., Pisacane, P. I., Sliwkowski, M. X., Forbes, J. T., and Arteaga, C. L. Inhibition of HER2/neu (erbB-2) and mitogen-activated protein kinases enhances tamoxifen action against HER2-overexpressing, tamoxifen-resistant breast cancer cells. Cancer Res, 60: 5887-5894, 2000.

262.   Dowsett, M. Overexpression of HER-2 as a resistance mechanism to hormonal therapy for breast cancer. Endocr Relat Cancer, 8: 191-195, 2001.

263.   Lipton, A., Leitzel, K., and Ali, S. M.

264.   Chen, X., Yeung, T. K., and Wang, Z. Enhanced drug resistance in cells coexpressing ErbB2 with EGF receptor or ErbB3. Biochem Biophys Res Commun, 277: 757-763, 2000.

265.   Bunone, G., Briand, P. A., Miksicek, R. J., and Picard, D. Activation of the unliganded estrogen receptor by EGF involves the MAP kinase pathway and direct phosphorylation. Embo J, 15: 2174-2183, 1996.

266.   Kato, S., Endoh, H., Masuhiro, Y., Kitamoto, T., Uchiyama, S., Sasaki, H., Masushige, S., Gotoh, Y., Nishida, E., and Kawashima, H. Activation of the estrogen receptor through phosphorylation by mitogen-activated protein kinase. Science, 270: 1491-1494, 1995.

267.   Campbell, R. A., Bhat-Nakshatri, P., Patel, N. M., Constantinidou, D., Ali, S., and Nakshatri, H. Phosphatidylinositol 3-kinase/AKT-mediated activation of estrogen receptor alpha: a new model for anti-estrogen resistance. J Biol Chem, 276: 9817-9824, 2001.

268.   De Laurentis, M., Bianco, A. R., and Placido, S.

269.   Hu, J. C. and Mokbel, K. Does c-erbB2/HER2 overexpression predict adjuvant tamoxifen failure in patients with early breast cancer? Eur J Surg Oncol, 27: 335-337, 2001.

270.   Love, R. R., Duc, N. B., Havighurst, T. C., Mohsin, S. K., Zhang, Q., DeMets, D. L., and Allred, D. C. Her-2/neu overexpression and response to

oophorectomy plus tamoxifen adjuvant therapy in estrogen receptor-positive premenopausal women with operable breast cancer. J Clin Oncol, 21: 453-457, 2003.

271. Ellis, M. J., Coop, A., Singh, B., Mauriac, L., Llombert-Cussac, A., Janicke, F., Miller, W. R., Evans, D. B., Dugan, M., Brady, C., Quebe-Fehling, E., and Borgs, M. Letrozole is more effective neoadjuvant endocrine therapy than tamoxifen for ErbB1- and/or ErbB-2-positive, estrogen receptor-positive primary breast cancer: evidence from a phase III randomized trial. J Clin Oncol, 19: 3808-3816, 2001.

272. Davol, P. A., Bagdasaryan, R., Elfenbein, G. J., Maizel, A. L., and Frackelton, A. R., Jr. Shc proteins are strong, independent prognostic markers for both node-negative and node-positive primary breast cancer. Cancer Res, 63: 6772-6783, 2003.

273. Barlow, R. E., Bartholomew, D. J., Bremmer, J. M., and Brunk, H. D. Statistical interference under order restrictions. London: John Wiley and Sons, Inc., 1972.

274. Breimann, L., Friedman, J. H., Olsen, R. A., and Stone, C. J. Classifiation and Regression Trees. Belmont: Wadsworth International Group, 1984.

275. Mantel, N. Evaluation of survival data and two new rank order statistics arising in its consideration. Cancer Chemother Rep, 50: 163-170, 1966.

276. Laemmli, U. K. Cleavage of structural proteins during the assembly of the head of bacteriophage T4. Nature, 227: 680-685, 1970.

Appendix 1. Schemes of sandwich CLISA assays

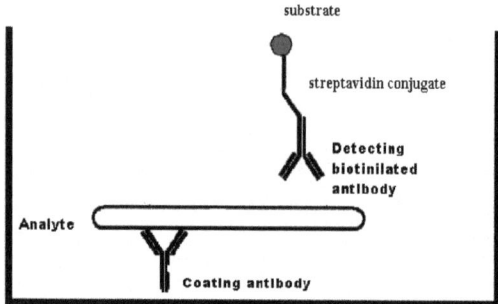

Figure 1. General layout of sandwich CLISA assays.

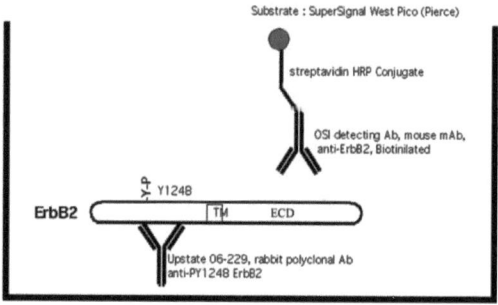

Figure 2. The layout of the immunoassay for P-Y1248 ErbB2 detection

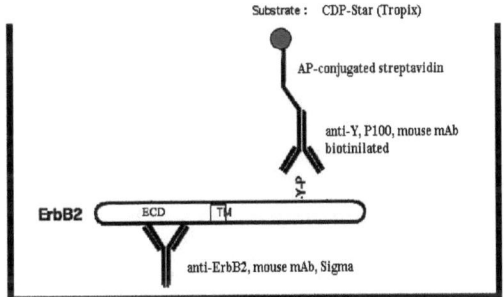

Figure 3. The layout of the immunoassay for pan-P-Y ErbB2 detection

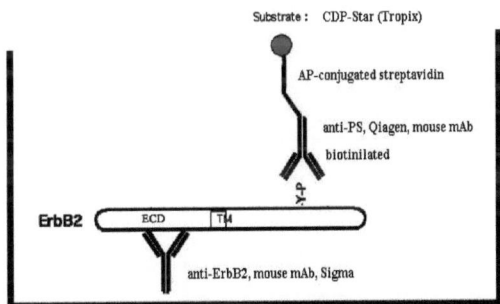

Figure 4. The layout of the immunoassay for pan-P-S ErbB2 detection

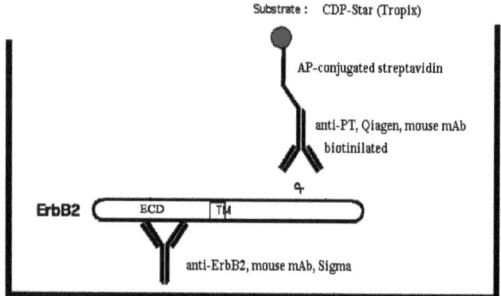

Figure 5. The layout of the immunoassay for pan-P-T ErbB2 detection

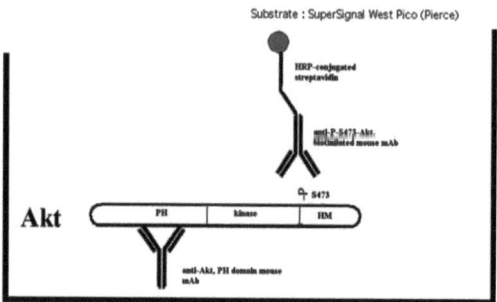

Figure 6. The layout of the immunoassay for P-S473 Akt detection

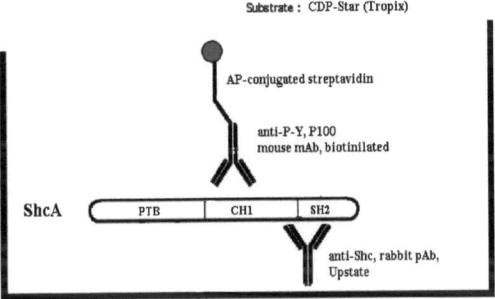

Figure 7. The layout of the immunoassay for pan-P-Y ShcA detection

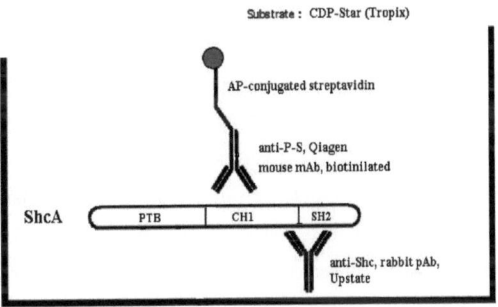

Figure 8. The layout of the immunoassay for pan-P-S ShcA detection

Appendix 2. PKCε in prostate cancer

Experiments were performed during 8.1999-5.2000 in Department of Anatomy and Cell Biology at East Carolina School of Medicine, under supervision by Prof. David M. Terrian, in collaboration with Dr. Daqing Wu.

Introduction.

Prostate cancer is the most frequently diagnosed cancer in men, accounting for 30% of all cancers, and is second leading cause of cancer deaths in men after lung cancer. In 2003, an estimated 220,900 American men were diagnosed with prostate cancer, and approximately 28,900 men died from the disease (1). Current methods of diagnosis including screening for high serum levels of prostate-specific antigen (PSA) and pathological grading of prostate biopsies cannot precisely distinguish between clinically aggressive and clinically indolent forms of prostate cancer. Therefore, far too many men are treated for the disease. Thus, there is an urgent need to identify biomarkers that distinguish the clinically aggressive forms of the tumor from the clinically indolent ones.

Prostatic epithelia normally depend on a functional androgen receptor (AR) signaling pathway for survival and undergo apoptosis in response to androgen ablation therapy (2). This accounts for the clinical regression such treatments initially produce among prostate cancer patients treated by endocrine therapies. However, a relapse of tumor growth is common, and these recurrent tumors are androgen-independent, highly metastatic, and hardly responsive to chemotherapy (3). Thus, oncogenic proteins that actively maintain the growth and survival of prostate cancer cells after androgen ablation make evident targets for the treatment of advanced prostate cancer. Many laboratories have attempted to identify these proteins through comprehensive analyses of differential gene expression between androgen-dependent and androgen-independent prostate cancer cell lines and prostate cancer biopsies (4, 5). Nevertheless, there is currently no direct evidence to support the concept of a dominant oncogene in recurrent prostate cancer.

The role of the ErbB2 receptor remains uncertain in the pathogenesis and progression of human prostate cancer. Several studies have reported widely divergent rates for ErbB2 expression in primary prostate tumors, probably owing to significant methodological differences in the studies (6, 7). Few data exist about the frequency of ErbB2 protein overexpression and gene amplification in androgen-independent prostate cancer, although xenograft models suggest ErbB2 expression may be up-regulated in the transition from androgen-dependent to androgen-independent disease (8). It has been shown that ErbB2 could induce AR transactivation at a low androgen level in two different prostate cancer cell lines, LNCaP and DU145 (9). Furthermore, transfection of ErbB2 in LNCaP resulted in higher expression of prostate-specific antigen (PSA) and higher proliferating rate. The data also suggested the ErbB2 could potentiate the expression of AR target genes through the MAP kinase pathway, possibly through direct AR phosphorylation.

PKCε is a member of the AGC family of Ser/Thr protein kinases that is known to have oncogenic potential (10) and to be associated with the progression of many cancers (11-14). Although there is recent evidence that PKCε expression is elevated in tissue biopsies collected from prostate cancer patients (15), the role of this isozyme in the progression to androgen independence has not been investigated. The activation of PKCα and PKCδ induces apoptosis in LNCaP cells, an intensively studied androgen-sensitive prostate cancer cell line, but not androgen-independent (DU145 and PC3) prostate cancer cell lines (16-18). This finding indicates that at least some members of this gene family are capable of differentially regulating the growth and survival of prostate cancer cells. Given the reciprocal functions of PKC isozymes in various cell types (10) and the oncogenic activity of PKCε, we hypothesized that this isozyme may oppose the proapoptotic influence of PKCα and PKCδ in prostate cancer.

In the present study, investigation of human prostate cancer cell lines indicated a relationship between PKCε expression and androgen independence. To better understand whether the expression of PKCε could be of functional importance in prostate cancer progression, we stably transfected LNCaP cells with a retroviral vector containing PKCε cDNA. This analysis revealed that PKCε overexpression was sufficient to transform LNCaP cells into an androgen-independent variant that rapidly initiated tumor growth in both intact and castrated male nude mice. The investigation also revealed that PKCε could potentiate androgen independence through the MAP kinase pathway. This study provides data demonstrating that PKCε expression may contribute to recurrent tumor growth in the absence of testicular androgens.

Materials and Methods.

Materials. [^{3}H]Thymidine (5 Ci/mmol) was obtained from Amersham (Piscataway, NJ, USA). PD098059 and LY294002 were from Sigma (St. Louis, MO, USA). Lipofectin and G418 were from Life Technologies (Carlsbad, CA, USA).

Cell Lines and Culture Conditions. LNCaP cell line was obtained from American Type Culture Collection (Manassas, VA; ATCC CRL-1740). All cell culture reagents were purchased from Invitrogen (Rockville, MD). LNCaP cells were maintained in culture in RPMI 1640 containing 2 mM L-glutamine, 10 mM HEPES, 1 mM sodium pyruvate, 4.5 g/l glucose, and 1.5 g/l sodium bicarbonate and supplemented with 10% FBS and 100 units/ml penicillin and 100 mg/ml streptomycin. Where indicated, cells were cultured in serum-free medium or in medium in which CS-FBS (Hyclone, Logan, UT) was substituted for untreated FBS. Cellular proliferation was assessed by the number of viable cells, which were counted, in triplicate, using a hemacytometer and trypan blue staining. Each assay was performed in triplicate in at least three independent experiments.

Generation of Overexpressing Cell Lines- LNCaP cells were infected with pLXSN recombinant retrovirus (LNCVa) or pLXSN harboring the gene for PKCε (LNW) Plasmid pLXSN constructs were transfected into the amphotrophic packaging line PA317 using Lipofectin and selecting with 400 μg/ml G418. The titer of the resulting retrovirus was amplified by sequential passage between the ψ2 and PA317 packaging cell lines before infection of LNCaP cells as described in Ref. 19. Stably expressing cells were selected and subcloned by limiting dilution (20) in 500 μg/ml G418, and resultant subclones were then screened for PKCε protein expression by Western blot analyses and representative clones were used for further studies.

Assessment of in Vivo Tumor Growth. Intact and surgically castrated nude male mice (NU/NU-*nu*BR) were purchased from Charles River Laboratories (Wilmington, MA) and inoculated subcutaneously, into the dorsal flanks left and right of the midline, with 1×10^6 cells suspended in 250 µl of PBS/site and routinely inspected for tumor growth and morbidity for up to 10 weeks. Cell cultures used in these studies were free of *Mycoplasma* contamination. Solid tumor volumes were calculated by the formula: length x width x depth x 0.5236.

[^{3}H]Thymidine **Incorporation Assays**—Proliferation was evaluated by [^{3}H]thymidine incorporation (at time 24 h). Cells (2×10^5/dish) were labeled for 12 h with [^{3}H]thymidine (0.5 µCi/ml) before harvesting and then washed three times with PBS; after addition of 3% perchloric acid, the acid-precipitable material was dissolved overnight in 1 N NaOH, 1% SDS and counted by liquid scintillation (Packard Tricarb 4530 counter).

Data Analysis. Values shown are representative of three or more experiments, unless otherwise specified, and treatment effects were evaluated using a two-sided Student's *t* test. Errors are SEs of averaged results, and values of $P < 0.05$ were taken as a significant difference between means.

Results.

PKCε Causes Androgen-independent Growth and Tumorigenicity. To determine whether signals transduced through PKCε had the potential to contribute to the androgen-idependent progression of prostate cancer, LNCaP cells overexpressing PKCε were established using the pLXSN retroviral vector. The pooled population of PKCε overexpressing LNCaP cells (LNW) were selected for by their collective resistance to G418, and a representative subclones (LNWa, LNWc) were isolated from this pool of transfectants by limiting dilution and maintained in culture. The pLXSN vector control line was called LNVa. LNWa, LNWc cells expressed equivalent levels of PKCε and, compared with parental and vector controls, the catalytic activity of PKCε was increased (not shown).

When cultured in complete medium, LNCaP cells maintained a functional androgen receptor signaling pathway and gradually became arrested in G_1 upon androgen removal (not shown). The overexpression of PKCε dramatically altered this phenotypic response to androgen withdrawal and enabled LNCaP cells to proliferate in the absence of androgens (Fig. 1). LNCaP cells overexpressing PKCε (LNW) remained responsive to the growth-promoting effects of DHT (Fig. 2). These results prompted investigation into the effects of PKCε overexpression on the tumorigenicity of LNCaP cells in the absence and presence of testicular androgens

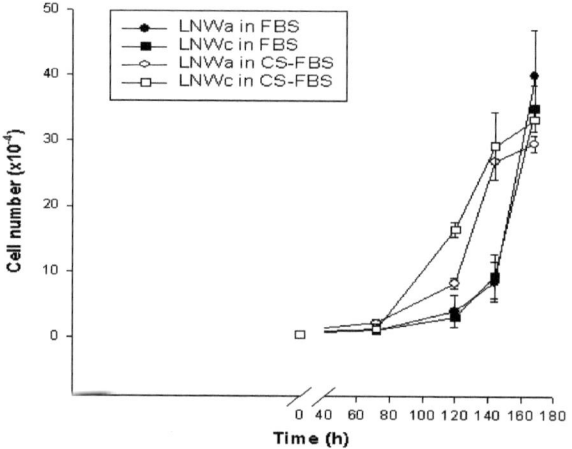

Figure 1. Growth of LNWa (○), LNWc (□) cells in charcoal treated medium (CS-FBS) and LNWa (●), and LNWc (■) cells in normal medium (FBS). Cell proliferation was measured by counting the total number of viable cells/plate by trypan blue exclusion using the hemacytometer. Data are the means of triplicate determinations in three independent experiments.

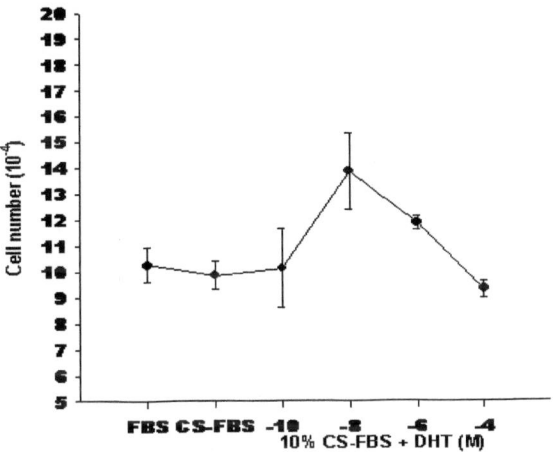

Figure 2. Growth-promoting effects of increasing concentrations of DHT on LNW cells incubated at 37°C for 3 days in CS-FBS. Cells were serum starved overnight before the introduction of DHT, and cell proliferation was measured by counting the total number of viable cells/plate by trypan blue exclusion. Data are the means of triplicate determinations in three independent experiments.

PKCε overexpressing cells rapidly initiate tumor growth in nude mice. Mycoplasma-free LNWa and LNWC cells (1×10^6/site) were injected subcutaneously into intact and castrated male nude mice. Matrigel was not used as an adjunct in any of these experiments, and all cells were injected alone as a suspension in PBS. Within 3–4 weeks, tumors appeared with a take rate of 100%, and the onset of tumor growth was more rapid in castrated animals (Table 1). No tumors formed during a 10-week observation period when an equal number of LNCaP cells were injected without Matrigel into intact and castrated male nude mice (Table 1).

Tumorigenicity in Nude Mice

Clone	% Tumor Formation	
	Intact	Castrated
LNCaP	0	0
LNWc	100	100
LNWa	100	100

Table 1. Intact and castrated male nude mice were injected subcutaneously with 1×10^6 LNCaP, LNWa or LNWc cells suspended in 0.25 ml of PBS. Data are the percentage of mice, which formed tumors for each group of eight mice.

Mitogen activated protein kinase (MAPK) is the downstream mediator PKCε mitogenic signal, leading to androgen-independence. PKCε has been implicated in a receptor tyrosine kinase (RTK) downstream signaling both via PI3K and Ras pathways (Fig. 3). ErbB2 is an RTK known to affect prostate cancer growth as well as androgen-independence via Ras/MAPK pathway. To determine whether MAPK cascade plays a role in androgen-idependent progression of prostate cancer mediated through PKCε LNW cells were treated with MEK-1 inhibitor PD098059. In order to rule out a possible direct effect of MAPK on proliferation of LNW cells, the difference between cells grown in FBS and CS-FBS was investigated.

A significant reduction in proliferation of cells treated with PD098059 was found between cells grown in CS-FBS (Figure 4), whereas the difference between cells grown in FBS was less prominent and statistically nonsignificant.

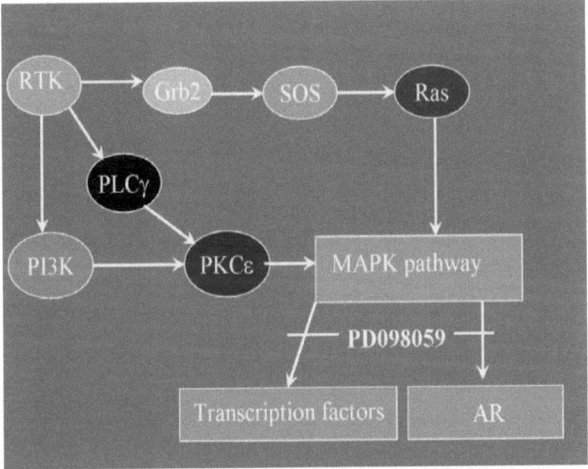

Figure 3. A scheme, depicting the role of PKCε in signaling through receptor tyrosine kinases and the possible effects of MEK1 inhibition.

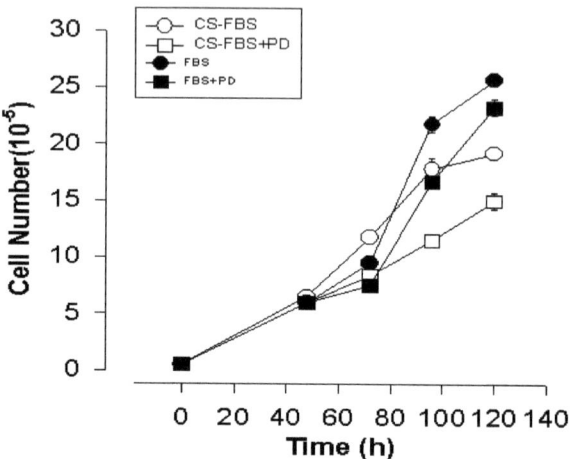

Figure 4. Growth of untreated (○), and PD098059 treated (□) cells in charcoal treated medium (CS-FBS) and untreated (●), and PD098059 treated (■) cells in normal medium (FBS). Cell proliferation was measured by counting the total number of viable cells/plate by trypan blue exclusion using the hemacytometer. Data are the means of triplicate determinations in three independent experiments.

PKCε is not sufficient to mediate the mitogenic signal downstream of PI3K. PKCε has been implicated in signaling via PI3K pathway, as a downstream mediator of mitogenic effects of PI3K. In order to determine whether overexpression of PKCε is sufficient in overcoming the effects of PI3K inhibition, LNWa and LNV cells were treated with LY294002, a specific inhibitor of PI3K, and then proliferation was analyzed using [3H]Thymidine incorporation assay. No difference was observed between proliferation of PKCε – overexpressing cells and vector control (Fig. 5).

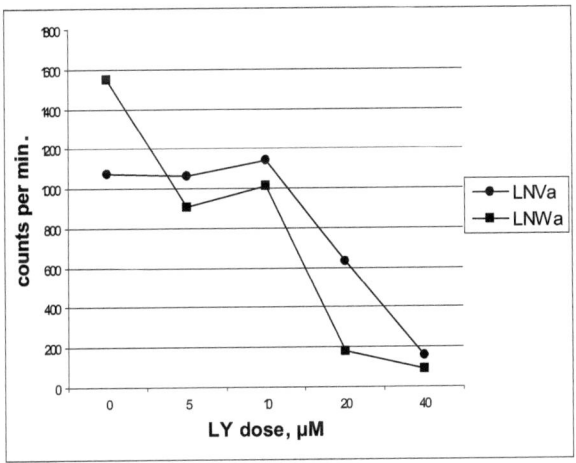

Figure 5. Proliferation of LNVa (●), and LNWa (■) cells treated with increasing doses of LY294002. Cell proliferation was measured by [3H]Thymidine incorporation in scintillation counter. Data are the means of triplicate measurements.

Discussion.

The major finding of the study was that PKCε is an oncogenic protein with the potential to induce androgen-independent growth of LNCaP tumors in both intact and castrated animals. These gene transfer experiments demonstrate that PKCε overexpression transforms LNCaP cells into androgen-idependent tumor cells that recapitulate many hallmark features of recurrent prostate cancer. The overexpression of PKCε leads to an uncontrolled and accelerated proliferation of LNCaP cells. The analysis also revealed that PKCε potentiates androgen independence, at least partially, through the MAP kinase pathway, most probably via Raf-1 activation. Although there is solid evidence that Raf-1 is a downstream target of PKCε, Ras/Raf induction alone is insufficient to promote the androgen-independent proliferation of LNCaP cells (21-22). Therefore, PKCε must signal to additional downstream targets to overcome the growth-regulatory signals that normally control the cell cycle progression of LNCaP cells. Although PI3K pathway is a major survival pathway of LNCaP cells, PKCε does not seem to be implicated in a downstream signaling of PI3K, promoting proliferation.

It is of note that multiple oncoproteins, and PKC isozymes, have been overexpressed in LNCaP cells without producing the phenotype of LNW cells. Although many important issues need to be further investigated, this study demonstrates that PKCε may play an important role in the progression to androgen independence in human prostate cancers.

References.

1. American Cancer Society. Cancer Facts & Figures 2003. Atlanta, GA: American Cancer Society.
2. Lieberman, R. Androgen deprivation therapy for prostate cancer chemoprevention: current status and future directions for agent development. Urology, 58: 83-90, 2001.
3. Oh, W. K. and Kantoff, P. W. Management of hormone refractory prostate cancer: current standards and future prospects. J Urol, 160: 1220-1229, 1998.
4. Gregory, C. W., Hamil, K. G., Kim, D., Hall, S. H., Pretlow, T. G., Mohler, J. L., and French, F. S. Androgen receptor expression in androgen-independent prostate cancer is associated with increased expression of androgen-regulated genes. Cancer Res, 58: 5718-5724, 1998.
5. Bubendorf, L., Kononen, J., Koivisto, P., Schraml, P., Moch, H., Gasser, T. C., Willi, N., Mihatsch, M. J., Sauter, G., and Kallioniemi, O. P. Survey of gene amplifications during prostate cancer progression by high-throughout fluorescence in situ hybridization on tissue microarrays. Cancer Res, 59: 803-806, 1999.
6. Ross, J. S., Sheehan, C. E., Hayner-Buchan, A. M., Ambros, R. A., Kallakury, B. V., Kaufman, R. P., Jr., Fisher, H. A., Rifkin, M. D., and Muraca, P. J. Prognostic significance of HER-2/neu gene amplification status by fluorescence in situ hybridization of prostate carcinoma. Cancer, 79: 2162-2170, 1997.
7. Kuhn, E. J., Kurnot, R. A., Sesterhenn, I. A., Chang, E. H., and Moul, J. W. Expression of the c-ErbB2 (HER-2/neu) oncoprotein in human prostatic carcinoma. J Urol, 150: 1427-1433, 1993.
8. Klein, K. A., Reiter, R. E., Redula, J., Moradi, H., Zhu, X. L., Brothman, A. R., Lamb, D. J., Marcelli, M., Belldegrun, A., Witte, O. N., and Sawyers, C. L. Progression of metastatic human prostate cancer to androgen independence in immunodeficient SCID mice. Nat Med, 3: 402-408, 1997.

9. Yeh, S., Lin, H. K., Kang, H. Y., Thin, T. H., Lin, M. F., and Chang, C. From HER2/Neu signal cascade to androgen receptor and its coactivators: a novel pathway by induction of androgen target genes through MAP kinase in prostate cancer cells. Proc Natl Acad Sci U S A, 96: 5458-5463, 1999.

10. Mischak, H., Goodnight, J. A., Kolch, W., Martiny-Baron, G., Schaechtle, C., Kazanietz, M. G., Blumberg, P. M., Pierce, J. H., and Mushinski, J. F. Overexpression of protein kinase C-delta and -epsilon in NIH 3T3 cells induces opposite effects on growth, morphology, anchorage dependence, and tumorigenicity. J Biol Chem, 268: 6090-6096, 1993.

11. Sharif, T. R. and Sharif, M. Overexpression of protein kinase C epsilon in astroglial brain tumor derived cell lines and primary tumor samples. Int J Oncol, 15: 237-243, 1999.

12. Perletti, G. P., Concari, P., Brusaferri, S., Marras, E., Piccinini, F., and Tashjian, A. H., Jr. Protein kinase Cepsilon is oncogenic in colon epithelial cells by interaction with the ras signal transduction pathway. Oncogene, 16: 3345-3348, 1998.

13. Knauf, J. A., Elisei, R., Mochly-Rosen, D., Liron, T., Chen, X. N., Gonsky, R., Korenberg, J. R., and Fagin, J. A. Involvement of protein kinase Cepsilon (PKCepsilon) in thyroid cell death. A truncated chimeric PKCepsilon cloned from a thyroid cancer cell line protects thyroid cells from apoptosis. J Biol Chem, 274: 23414-23425, 1999.

14. Lavie, Y., Zhang, Z. C., Cao, H. T., Han, T. Y., Jones, R. C., Liu, Y. Y., Jarman, M., Hardcastle, I. R., Giuliano, A. E., and Cabot, M. C. Tamoxifen induces selective membrane association of protein kinase C epsilon in MCF-7 human breast cancer cells. Int J Cancer, 77: 928-932, 1998.

15. Cornford, P., Evans, J., Dodson, A., Parsons, K., Woolfenden, A., Neoptolemos, J., and Foster, C. S. Protein kinase C isoenzyme patterns characteristically modulated in early prostate cancer. Am J Pathol, 154: 137-144, 1999.

16. Henttu, P. and Vihko, P. The protein kinase C activator, phorbol ester, elicits disparate functional responses in androgen-sensitive and androgen-independent human prostatic cancer cells. Biochem Biophys Res Commun, 244: 167-171, 1998.

17. Fujii, T., Garcia-Bermejo, M. L., Bernabo, J. L., Caamano, J., Ohba, M., Kuroki, T., Li, L., Yuspa, S. H., and Kazanietz, M. G. Involvement of protein kinase C delta (PKCdelta) in phorbol ester-induced apoptosis in LNCaP prostate cancer cells. Lack of proteolytic cleavage of PKCdelta. J Biol Chem, 275: 7574-7582, 2000.

18. Garzotto, M., White-Jones, M., Jiang, Y., Ehleiter, D., Liao, W. C., Haimovitz-Friedman, A., Fuks, Z., and Kolesnick, R. 12-O-tetradecanoylphorbol-13-acetate-induced apoptosis in LNCaP cells is mediated through ceramide synthase. Cancer Res, 58: 2260-2264, 1998.

19. Miller, A. D. and Rosman, G. J. Improved retroviral vectors for gene transfer and expression. Biotechniques, 7: 980-982, 984-986, 989-990, 1989.

20. Kiley, S. C., Adams, P. D., and Parker, P. J. Cloning and characterization of phorbol ester differentiation-resistant U937 cell variants. Cell Growth Differ, 8: 221-230, 1997.

21. Ravi, R. K., McMahon, M., Yangang, Z., Williams, J. R., Dillehay, L. E., Nelkin, B. D., and Mabry, M. Raf-1-induced cell cycle arrest in LNCaP human prostate cancer cells. J Cell Biochem, 72: 458-469, 1999.

22. Marais, R., Light, Y., Mason, C., Paterson, H., Olson, M. F., and Marshall, C. J. Requirement of Ras-GTP-Raf complexes for activation of Raf-1 by protein kinase C. Science, 280: 109-112, 1998.

Jonas Cicenas
Curriculum Vitae

Date and place of birth: December 17, 1974; Klaipeda, Lithuania.

Citizenship: Lithuania (EU).

Marital Status: Married (Ernesta Ciceniene, MSc in biology)

Education:

1982-1993:Klaipeda School No. 19;

1993-1997 and 1997-1999: Faculty of Natural Sciences, Vilnius Pedagogical University(biology);

1997: Bachelor's degree in natural sciences;

1999: Master's degree in natural sciences.

1999-2000: Dept. Anatomy and Cell Biology, East Carolina University School of Medicine.

2002-2004: University of Basel

Research and technical experience:

1996-1997: Institute of Immunology, Lithuania (undergraduate student);

1997-1999: Institute of Biochemistry, Lithuania (laboratory assistant);

1999-2000: Dept. Anatomy and Cell Biology, East Carolina University School of Medicine, Greenville, NC, USA (PhD student).

2002-2003: Molekulare Tumorbiologie, Department Forshung, Kantonspital Basel and 2003-2004: Stiftung Tumorbank Basel (PhD student)

Publications

Weissenstein U*, Schneider MJ*, Pawlak M*, Cicenas J*, Eppenberger-Castori S, Oroszlan P, Ehret S, Geurts-Moespot A, Sweep FC, Eppenberger U. **Protein chip based miniaturized assay for the simultaneous quantitative monitoring of cancer biomarkers in tissue extracts.**- Proteomics. 2006 Mar;6(5):1427-36* contributed equally.

Cicenas J, Urban P, Kung W, Vuaroqueaux V, Labuhn M, Wight E, Eppenberger U, Eppenberger-Castori S. **Phosphorylation of tyrosine 1248-ERBB2 measured by chemiluminescence-linked immunoassay is an independent predictor of poor prognosis in primary breast cancer patients.**- Eur J Cancer. 2006 Mar;42(5):636-45.

Cicenas J, Urban P, Vuaroqueaux V, Labuhn M, Kung W, Wight E, Mayhew M, Eppenberger U, Eppenberger-Castori S. **Increased level of phosphorylated akt measured by chemiluminescence-linked immunosorbent assay is a predictor of poor prognosis in primary breast cancer overexpressing ErbB-2.**-Breast Cancer Res. 2005;7(4):R394-401. Epub 2005 Mar 24.

Die VDM Verlagsservicegesellschaft sucht für wissenschaftliche Verlage abgeschlossene und herausragende

Dissertationen, Habilitationen, Diplomarbeiten, Master Theses, Magisterarbeiten usw.

für die kostenlose Publikation als Fachbuch.

Sie verfügen über eine Arbeit, die hohen inhaltlichen und formalen Ansprüchen genügt, und haben Interesse an einer honorarvergüteten Publikation?

Dann senden Sie bitte erste Informationen über sich und Ihre Arbeit per Email an *info@vdm-vsg.de*.

Sie erhalten kurzfristig unser Feedback!

VDM Verlagsservicegesellschaft mbH
Dudweiler Landstr. 99　　　　　Telefon +49 681 3720 174
D - 66123 Saarbrücken　　　　　Fax　　　+49 681 3720 1749
www.vdm-vsg.de

Die VDM Verlagsservicegesellschaft mbH vertritt

Printed by Books on Demand GmbH, Norderstedt / Germany